零基础轻松学
Java

张洪波 丁卫颖 郑 铮 编著

机械工业出版社
China Machine Press

图书在版编目（CIP）数据

零基础轻松学Java / 张洪波，丁卫颖，郑铮编著. - 北京：机械工业出版社，2018.10
ISBN 978-7-111-61130-1

I. ①零… II. ①张… ②丁… ③郑… III. ①JAVA语言 – 程序设计 IV. ①TP312.8

中国版本图书馆CIP数据核字（2018）第233950号

本书系统地介绍了Java程序设计的基础知识、开发环境与开发工具。全书共12章，内容包括Java语言概述、Java语言的基本语法、程序流程控制语句、面向对象编程基础、异常的概念和处理、线程处理机制、集合框架、Java输入输出/（IO）处理、JDBC数据库编程、Java网络编程和Swing程序设计，最后介绍Java常用的类库和开发Java程序的实战演练，包括记事本工具、网络通信工具和在线相册的开发项目。另外，每章还安排了练习题和编程练习，供读者巩固知识，提升编程技能。

本书从初学者的角度出发，以丰富的实例、通俗易懂的语言、简单的图示，详细介绍Java开发中重点用到的多种技术，使读者快速掌握Java程序设计的方法。本书适合学习Java编程的初学者使用，也可作为普通高等院校计算机及相关专业Java程序设计的教材。

零基础轻松学Java

出版发行：机械工业出版社（北京市西城区百万庄大街22号　邮政编码：100037）			
责任编辑：夏非彼　迟振春		责任校对：闫秀华	
印　　刷：中国电影出版社印刷厂		版　次：2018年11月第1版第1次印刷	
开　　本：188mm×260mm　1/16		印　张：20	
书　　号：ISBN 978-7-111-61130-1		定　价：69.00元	

凡购本书，如有缺页、倒页、脱页，由本社发行部调换
客服热线：（010）88379426　88361066　　　　　　投稿热线：（010）88379604
购书热线：（010）68326294　88379649　68995259　　读者信箱：hzit@hzbook.com

版权所有·侵权必究
封底无防伪标均为盗版
本书法律顾问：北京大成律师事务所　韩光/邹晓东

前　言

　　Java 是 Sun 公司推出的一种程序设计语言，拥有面向对象、跨平台、分布式、高性能、可移植等优点和特性，是目前广泛使用的编程语言之一。Java 主要有 Java SE（Java 标准版本）、Java EE（Java 企业版本）和 Java ME（Java 移动电子设备版本）三个版本。其中，Java SE 是 Java 语言的标准版，包含 Java 基本语法、面向对象程序设计、多线程、数据集合、输入和输出、Swing 程序设计、网络编程及数据库操作等。

　　本书通过通俗易懂的语言和实用生动的例子，系统地介绍了 Java SE 程序设计的基础知识、开发环境与开发工具，并且在每一章的后面都提供习题，方便读者检查学习效果，还提供了部分上机实验，使读者快速掌握 Java SE 程序的开发技能。

　　全书共分 12 章，具体内容如下：

　　第 1 章主要介绍 Java 技术的相关概念及如何开发 Java 程序，包括 Java 语言的概述、Java 开发环境的搭建、开发工具 Ecplise 的使用。

　　第 2 章主要介绍 Java 语言的基础知识，包括数据类型、运算符、控制流程、数组和函数等。

　　第 3 章主要介绍面向对象编程的基础，包括 Java 语言中的类、对象、包、接口、类的继承、多态和封装等。

　　第 4 章介绍异常处理机制，包括异常类及分类、异常处理机制、自定义异常等。

　　第 5 章介绍线程技术，包括线程的创建、状态、调度、优先级和线程同步等。

　　第 6 章主要介绍 Java 程序设计中的集合类，包括 List 集合、Set 集合、Map 集合等。

　　第 7 章主要介绍 Java 的输入 / 输出功能，包括文件操作类、字节流、字符流和对象序列化等。

　　第 8 章主要介绍 Java 的数据库操作，包括 JDBC 概述、JDBC 中常用的接口、连接数据库、操作数据库和应用 JDBC 事务。

　　第 9 章主要介绍 Java 的网络编程，通过聊天程序讲解网络编程的相关技术。

　　第 10 章主要介绍 Java 的图形用户界面编程，讲解 Swing 的编程技术和实现。

第 11 章主要介绍 Java 常用的类库，包括 StringBuffer 类、Runtime 类、System 类、Math 类、Random 类等。

第 12 章通过实战演练让读者了解和掌握 Java 程序开发的过程。

本书资源可以登录机械工业出版社华章公司的网站（www.hzbook.com）下载，搜索到本书，然后在页面上的"资源下载"模块下载即可。

本书由唐山师范学院的张洪波、丁卫颖、郑铮老师共同编写完成，第 1~3、7、8、10、11 章由张洪波编写，第 4～6 章由丁卫颖编写，第 9 章和第 12 章由郑铮编写。同时，对参与文字录入及书中代码编写、调试工作的人员表示衷心的感谢！如果读者对本书有疑问或建议，可发送电子邮件至 booksaga@126.com。

因为要熟练掌握 Java 语言必须进行大量的上机练习，所以无论是否安排上机实验，读者都应该独立、认真地完成本书中的所有示例和实验。本书适合 Java 初学者和计算机及相关专业的学生使用。

由于编写时间仓促和作者水平有限，书中错误和不妥之处敬请读者批评指正。

编　者

2018 年 9 月

目 录

前 言

第 1 章 开始 Java 之旅 ··· 1
1.1 无处不在的 Java ··· 1
1.2 Java 为何受大家喜爱 ··· 2
1.3 Java 的目标 ··· 3
1.4 Java 开发环境的搭建 ··· 3
1.4.1 JDK 的下载与安装 ·· 3
1.4.2 Java 开发环境配置 ·· 5
1.5 Java 程序运行的原理 ··· 6
1.6 Java 开发工具 Eclipse ··· 6
1.6.1 Eclipse 的安装与启动 ·· 6
1.6.2 Eclipse 编写 Java 程序的流程 ··· 8
1.7 要点总结 ··· 12
1.8 练习题 ··· 12

第 2 章 Java 语言基础 ··· 13
2.1 Java 程序的基本组成 ··· 13
2.2 Java 语言的数据类型 ··· 16
2.2.1 整数类型 ·· 16
2.2.2 浮点类型 ·· 17
2.2.3 字符类型 ·· 17
2.2.4 布尔类型 ·· 17
2.2.5 基本数据类型的默认值 ·· 17
2.2.6 类型转换 ·· 17
2.3 运算符和表达式 ··· 19
2.3.1 赋值运算符 ·· 19
2.3.2 算术运算符 ·· 19
2.3.3 位运算符 ·· 20

2.3.4 关系运算符 ... 21
2.3.5 三元运算符 ... 21
2.3.6 运算符优先级 ... 22
2.4 流程控制语句 ... 22
2.4.1 选择语句 ... 22
2.4.2 循环语句 ... 27
2.5 数组与方法 ... 32
2.5.1 一维数组 ... 32
2.5.2 二维数组 ... 35
2.5.3 方法 ... 37
2.6 要点总结 ... 40
2.7 编程练习 ... 40

第 3 章 Java 面向对象编程 ... 41

3.1 理解面向对象 ... 41
3.1.1 基本概念 ... 41
3.1.2 基本特性 ... 42
3.2 类与对象 ... 42
3.2.1 类定义 ... 42
3.2.2 对象的创建及使用 ... 45
3.2.3 this 和 static 关键字 ... 48
3.2.4 内部类 ... 52
3.3 继承 ... 53
3.3.1 继承的语法和规则 ... 53
3.3.2 重载和覆盖 ... 55
3.3.3 super 关键字 ... 56
3.4 final 关键字 ... 57
3.4.1 final 变量 ... 57
3.4.2 final 方法 ... 57
3.4.3 final 类 ... 58
3.5 抽象类 ... 58
3.6 接口 ... 59
3.6.1 接口定义 ... 59
3.6.2 实现接口 ... 60
3.6.3 匿名内部类 ... 61
3.7 包及访问控制权限 ... 61

3.7.1 包的操作 ... 62
3.7.2 访问权限修饰符 ... 62
3.8 对象的多态性 ... 63
3.9 Object 类 ... 64
3.10 包装类 ... 66
3.10.1 基本数据类型转换为包装类 ... 67
3.10.2 字符串转换为包装类 ... 67
3.10.3 包装类转换为基本数据类型 ... 67
3.10.4 字符串转换为基本数据类型 ... 68
3.10.5 自动装箱和自动拆箱 ... 68
3.10.6 覆盖父类的方法 ... 69
3.11 String 类 ... 69
3.11.1 String 对象的实例化和内容比较 ... 69
3.11.2 String 类中的常用方法 ... 70
3.12 要点总结 ... 71
3.13 编程练习 ... 71

第 4 章 Java 异常 ... 73

4.1 Java 中的异常类及分类 ... 73
4.2 Java 异常处理机制 ... 75
4.2.1 捕获处理异常 ... 76
4.2.2 声明抛出异常 ... 80
4.3 自定义异常 ... 82
4.4 自定义异常的综合应用 ... 84
4.5 实例练习：异常的综合应用 ... 87
4.6 要点总结 ... 88
4.7 编程练习 ... 88

第 5 章 Java 线程 ... 90

5.1 多线程及线程简介 ... 90
5.2 线程的创建 ... 91
5.3 线程的状态 ... 95
5.4 线程的调度 ... 95
5.5 线程的优先级 ... 96
5.6 守护线程 ... 99
5.7 线程同步 ... 101
5.8 实例练习：线程综合应用 ... 106

5.9 要点总结 ... 109
5.10 练习题 ... 109
5.11 编程练习 ... 110

第 6 章 ... 111

Java 集合框架 ... 111

6.1 常用集合接口 ... 111
6.1.1 Collection 接口 ... 112
6.1.2 List 接口 ... 113
6.1.3 Set 接口 ... 114
6.1.4 Map 接口 ... 114
6.1.5 Map.Entry 接口 ... 116
6.1.6 Iterator 接口 ... 116
6.1.7 ListIterator 接口 ... 116

6.2 常用集合类 ... 117
6.2.1 ArrayList 类 ... 118
6.2.2 LinkedList 类 ... 121
6.2.3 HashSet 类 ... 125
6.2.4 HashMap 类 ... 127

6.3 实例练习：集合类的综合运用 ... 131
6.4 要点总结 ... 132
6.5 练习题 ... 132
6.6 编程练习 ... 133

第 7 章 Java IO ... 134

7.1 File 类 ... 134
7.2 RandomAccessFile 类 ... 136
7.3 字节流与字符流 ... 137
7.3.1 字节流 ... 138
7.3.2 字符流 ... 139
7.3.3 字节流与字符流的区别 ... 140
7.4 转换流 ... 141
7.5 打印流 ... 142
7.6 管道流 ... 143
7.7 BufferedReader 类和 BufferedWriter 类 ... 144
7.8 数据操作流 ... 145

7.9 对象流 147

7.10 Scanner 类 148

7.11 要点总结 148

7.12 编程练习 149

第 8 章 Java 数据库编程 150

8.1 JDBC 技术 150

    8.1.1 JDBC 技术简介 150

    8.1.2 JDBC 驱动程序 151

    8.1.3 JDBC 和 ODBC 与其他 API 的比较 152

8.2 结构化查询语言 152

    8.2.1 SQL 简介 153

    8.2.2 SELECT 语句 153

    8.2.3 更新记录 154

    8.2.4 聚集函数 155

8.3 JDBC 基本操作 156

    8.3.1 JDBC 操作步骤 156

    8.3.2 JDBC-ODBC 连接数据库 157

    8.3.3 JDBC 直接连接数据库 163

    8.3.4 JDBC 对数据库的更新操作 167

8.4 JDBC 高级操作 170

    8.4.1 PreparedStatemen 接口 170

    8.4.2 CallableStatement 接口 172

    8.4.3 事务处理 173

8.5 要点总结 173

8.6 练习题 173

8.7 编程练习 174

第 9 章 Java 网络编程 175

9.1 网络基础 175

    9.1.1 TCP/IP 网络模型 175

    9.1.2 IP 地址与 InetAddress 类 176

    9.1.3 套接字 176

9.2 UDP 协议网络程序 177

    9.2.1 概述 177

    9.2.2 DatagramPacket 类 177

    9.2.3 DatagramSocket 类 178

9.2.4 创建 UDP 服务器端程序 ... 179
9.2.5 创建 UDP 客户端程序 ... 180
9.3 TCP 协议网络程序 ... 181
9.3.1 概述 ... 181
9.3.2 Socket 类 ... 182
9.3.3 ServerSocket 类 ... 183
9.3.4 创建 TCP 服务器端程序 ... 184
9.3.5 创建 TCP 客户端程序 ... 185
9.4 HTTP 协议网络程序 ... 186
9.4.1 概述 ... 186
9.4.2 URL 类 ... 186
9.4.3 URLConnection 类 ... 188
9.5 综合实例：实现简单的 Web 服务器 ... 189
9.6 要点总结 ... 193
9.7 练习题 ... 193
9.8 编程练习 ... 194

第 10 章 Java 图形用户界面 ... 195

10.1 AWT 与 Swing 简介 ... 195
10.1.1 AWT 简介 ... 195
10.1.2 Swing 简介 ... 196
10.1.3 容器简介 ... 196
10.2 创建窗体 ... 197
10.3 标签组件：JLabel ... 198
10.4 按钮组件：JButton ... 202
10.5 JPanel 容器 ... 203
10.6 布局管理器 ... 204
10.6.1 FlowLayout ... 204
10.6.2 BorderLayout ... 205
10.6.3 GridLayout ... 206
10.6.4 CardLayout ... 207
10.7 文本组件：JTextComponent ... 208
10.7.1 单行文本框：JTextField ... 208
10.7.2 密码文本框：JPasswordField ... 210
10.7.3 多行文本框：JTextArea ... 211
10.8 事件处理 ... 212

  10.8.1 事件和监听器 ·· 212

  10.8.2 窗体事件 ·· 215

  10.8.3 动作事件及监听处理 ····································· 218

  10.8.4 键盘事件及监听处理 ····································· 219

  10.8.5 鼠标事件及监听处理 ····································· 221

  10.8.6 焦点事件及监听处理 ····································· 223

10.9 单选按钮组件：JRadioButton ······································ 224

10.10 复选框组件：JCheckBox ·· 227

10.11 列表框组件：JList ··· 229

10.12 下拉列表框：JComboBox ··· 230

10.13 菜单组件：JMenu 与 JMenuBar ································· 232

10.14 文件选择框组件：JFileChooser ··································· 233

10.15 要点总结 ··· 236

10.16 练习题 ·· 236

第 11 章 Java 常用类库 ·· 239

11.1 StringBuffer 类 ··· 239

11.2 Runtime 类 ··· 242

11.3 System 类 ·· 244

11.4 Math 类 ··· 244

11.5 Random 类 ·· 245

11.6 要点总结 ··· 245

11.7 练习题 ··· 245

第 12 章 Java 项目开发 ·· 248

12.1 软件开发过程 ··· 248

  12.1.1 需求 ·· 248

  12.1.2 分析设计 ··· 249

  12.1.3 实现和测试 ··· 249

12.2 项目实例：记事本工具的开发 ····································· 249

  12.2.1 需求分析设计 ·· 249

  12.2.2 实现和测试 ··· 250

12.3 项目实例：网络通信工具的开发 ································· 267

  12.3.1 需求分析设计 ·· 267

  12.3.2 实现和测试 ··· 268

12.4 项目实例：在线相册的开发 ··· 272

  12.4.1 需求分析设计 ·· 272

12.4.2 数据库设计 ……………………………………………………………… 273
12.4.3 开发数据库 JavaBean ………………………………………………… 276
12.4.4 实现和测试 ……………………………………………………………… 290
12.5 要点总结 …………………………………………………………………………… 308
12.6 编程练习 …………………………………………………………………………… 308

第 1 章
开始 Java 之旅

Java 是由 Sun 公司开发的一种应用于分布式网络环境的程序设计语言。Java 语言拥有跨平台的特性,其编译的程序能够运行在多种系统操作平台上,可以实现"一次编写,到处运行"。本章主要介绍 Java 语言的特点、目标、开发环境的搭建、运行原理及开发工具的使用。

1.1 无处不在的 Java

一般的初学者都认为 Java 是一种编程语言,实际上,Java 不仅是一种语言,更是一个平台。它还提供了开发类库、运行环境、部署环境等一系列的支持。

根据 Java 应用范围的不同,可以分为三个版本:Java SE、Java EE 和 Java ME。

- Java SE(Java Standard Edition)包含了标准的 JDK、开发工具、运行环境和类库,适合开发桌面应用程序和底层应用程序,同时也是 Java EE 的基础平台。
- Java EE(Java Enterprise Edition)采用标准化的模块组件为企业级应用提供了标准平台,简化了复杂的企业级编程。现在 Java EE 已经成为了一种软件架构和企业级开发的设计思想。
- Java ME(Java Micro Edition)包含高度优化精简的 Java 运行环境,主要用于开发具有有限的连接、内存和用户界面能力的设备应用程序,如移动电话(手机)、PDA(电子商务)、能够接入电缆服务的机顶盒或各种终端和其他消费电子产品。

目前无论是银行管理还是手机消费，从科学研究的巨型计算机到笔记本电脑，Java 的身影无处不在，已经成为行业内非常流行且时髦的编程技术。

1.2 Java 为何受大家喜爱

Java 语言具有简单、面向对象、分布式、解释器通用性、健壮、安全、可移植性、高效能、多线程、动态等语言特性。另外，还提供了丰富的类库，方便用户进行自定义操作。

1. 简单

Java 在设计上与 C++ 十分相近。Java 中删除了许多极少被使用、不容易理解和令人混淆的 C++ 功能，如运算符重载、多重继承等，增加了内存垃圾自动收集功能，关于内存的分配与释放是使 C 与 C++ 应用程序变得复杂的常见原因之一。因为 Java 的垃圾自动收集功能简化了程序设计工作，所以无论是经验丰富的 C++/C 程序员还是程序设计的初学者，学习 Java 都是非常容易的。

2. 面向对象

Java 语言以面向对象为基础。在 Java 语言中，不能在类外面定义单独的数据和函数，所有的对象都要派生于同一个基类，并共享其所有的功能。也就是说，Java 语言最外部的数据类型是对象，所有的元素都要通过类和对象来访问。

3. 分布式

由于 Java 中内置了 TCP/IP、HTTP、FTP 等协议，因此 Java 应用程序可以通过 URL 地址访问网络上的对象，访问方式与访问本地文件系统几乎完全相同。

4. 解释器通用性

Java 解释器能直接对 Java 字节码进行解释执行。经过编译生成的字节码可以在提供 Java 虚拟机的任何一个系统上解释运行，不需要额外存储。

5. 健壮

Java 能够检查程序在编译和运行时的错误。类型检查能帮助用户检查出许多在开发早期出现的错误，同时许多集成开发环境（IDE）的出现使编译和运行 Java 程序更加容易。

6. 安全

因为 Java 的设计目标是提供使用于网络/分布式运算环境，所以安全性问题自然是不容忽视的。Java 的验证技术是以公钥加密法为基础的。

7. 可移植性

Java 程序具有与体系结构无关的特性。这一特性使 Java 程序可以方便地移植到网络上不同的机器。同时 Java 的类库中也实现了针对不同平台的接口，使这些类库可以移植。

8. 高效能

虽然 Java 字节码是解释运行，但是经过仔细设计的字节码可以通过 JIT 技术转换为高效能的本机代码。

9. 多线程

Java 支持多线程编程。Java 运行时，系统在多线程同步方面具有成熟的解决方案。这使得程序设计者将更多的精力关注于程序实现的细节。

1.3 Java 的目标

Internet 的迅猛发展，使 Java 成为当前非常流行的网络编程语言。最初设计 Java 有以下几个目标：

（1）不依赖于特定的平台，一次编写，到处运行。

（2）完全面向对象。

（3）内置对计算机网络的支持。

（4）借鉴 C++ 优点，尽量简单易用。

1.4 Java 开发环境的搭建

本节介绍 Java 开发环境的搭建，主要包括 JDK 的下载与安装及配置 Java 开发环境。

1.4.1 JDK 的下载与安装

JDK（Java Development Kit）是 Java 的开发工具包，亦是 Java 开发者必须安装的软件环境。JDK 包含了 JRE 与开发 Java 程序所需的工具，如编译器、调试器、反编译器、文档生成器等。

JRE（Java Runtime Environment）是 Java 程序运行的必要环境，包含类库和 JVM（Java 虚拟机）。如果只运行 Java 程序，就没有必要安装 JDK，只安装 JRE 即可。

Sun 公司网站下载 JDK1.6 的地址为 http://java.sun.com/javase/downloads/index.html。

注意，Java 是跨平台的开发语言，根据平台的不同，要选择不同的 JDK，本书选择 Windows platform。在这里，JDK 又分为在线安装包和离线安装包两种，选择离线安装方式。

将下载的 JDK1.6 安装包保存到硬盘上，文件名为 jdk-6u2-windows-i586-p.exe，运行该文件并按照向导进行安装。关闭所有正在运行的程序，接受许可协议，设置 JDK 的安装路径及选择安装的组件对话框，如图 1-1 所示。

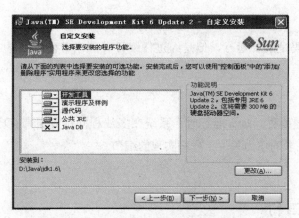

图 1-1 设置 JDK 的安装路径及选择安装的组件对话框

更改安装路径到 D:\Java\jdk1.6,选择要安装的组件。在安装过程中,定义 JRE 安装路径的提示对话框,更改路径到 D:\Java\jdk1.6。在弹出安装完成的提示对话框中,取消对"显示自述文件"复选框的勾选,单击"完成"按钮,即可完成 JDK 的安装。

安装目录路径如图 1-2 所示。

图 1-2 JDK 安装路径

主要目录和文件简介如下。

- bin 目录:开发工具,包括开发、运行、调试和文档生成的工具,主要是 *.exe 文件。
- lib 目录:类库,开发时需要的一些类库和文件。
- jre 目录:运行时环境,包括 Java 虚拟机、类库、辅助运行的支持文件。
- demo 目录:演示文件,附源代码的 Java 文件,演示了 Java 的一些功能。
- include 目录:C 语言头文件,支持 Java 本地方法调用的必要文件。
- src.zip 文件:Java 核心类源文件,感兴趣的读者可以解压后研究。

其中，bin 目录中的两个文件非常重要，在编程中经常使用。

- javac.exe：Java 编译器。
- java.exe：Java 解释器，调用 Java 虚拟机执行 Java 程序。

执行"开始"→"运行"→输入 cmd，如图 1-3 所示。

进入 DOS 命令行窗口，输入 Java –version，效果如图 1-4 所示，即为安装成功。

图 1-3 输入 cmd

图 1-4 测试 JDK 是否安装成功

1.4.2 Java 开发环境配置

安装完 JDK 后，需要设置环境变量及测试 JDK 配置是否成功。在 Windows XP 系统下的具体操作步骤如下：

步骤 01 在"我的电脑"上单击鼠标右键，在弹出的快捷菜单中选择"属性"命令。在打开的"系统属性"对话框中选择"高级"选项卡，单击"环境变量"按钮，打开"环境变量"对话框，单击针对所有用户的"系统变量"区域中的"新建"按钮。

步骤 02 弹出"新建系统变量"对话框，在"变量名"文本框中输入 JAVA_HOME，在"变量值"文本框中输入 JDK 的安装路径，单击"确定"按钮，如图 1-5 所示，完成环境变量 JAVA_HOME 的配置。

步骤 03 在"系统变量"区域中查看 path 变量，若不存在，则新建变量 PATH，否则选中该变量，单击"编辑"按钮，弹出"新建系统变量"对话框，在"变量值"文本框的起始位置添加 %JAVA_HOME%\bin，然后单击"确定"按钮。注意，不要漏掉最后的"；"符号。

步骤 04 在"系统变量"，区域中查看 classpath 变量，若不存在，则新建变量 classpath，单击

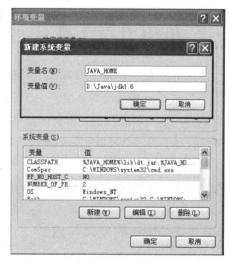

图 1-5 JAVA_HOME 环境变量的配置

"新建"按钮,弹出"新建系统变量"对话框,在"变量值"文本框中输入 %JAVA_HOME%\lib\dt.jar;%JAVA_HOME%\lib\tools.jar,然后单击"确定"按钮。

步骤 05 测试 JDK 是否能够在机器上运行。在 DOS 命令行窗口输入"javac",输出帮助信息即为配置正确。

1.5 Java 程序运行的原理

下面利用 Windows 记事本程序编写一个简单的 Java 文件,如图 1-6 所示。

将代码保存到 D 盘,并命名为 HelloWorld.java。在 DOS 命令行窗口编译源代码:Javac HelloWorld.java,编译正确生成 Hello.class 文件;Java 解释器解释执行 class 文件:Java Hello。

可由运行过程了解 Java 的运行原理,如图 1-7 所示。

图 1-6 Java 的开发过程

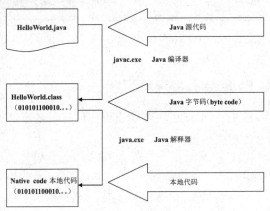

图 1-7 Java 程序运行原理

1.6 Java 开发工具 Eclipse

Eclipse 是一个基于 Java 的、开放源代码的、可扩展的应用开发平台,为编程人员提供了一流的 Java 集成开发环境。由于 Eclipse 是利用 Java 语言写成的,因此 Eclipse 支持跨平台操作,是一个成熟的、可扩展的体系结构。Eclipse 的价值还体现在为创建可扩展的开发环境提供了一个开发源代码的平台,这个平台允许任何人构建与环境或其他工具无缝集成的工具。而工具与 Eclipse 无缝集成的关键是插件,通过不断地集成各种插件,Eclipse 的功能也在不断扩展,以便支持各种不同的应用。

1.6.1 Eclipse 的安装与启动

安装 Eclipse 前需要先安装 JDK,关于 JDK 的安装与配置请参见 1.4 节中的内容。可以从 Eclipse 的官方网站（http://www.eclipse.org）下载当前最新的版本。本书使用的 Eclipse 版本是 3.5。

Eclipse 下载完成后，解压并运行文件，即完成对 Eclipse 的安装。

初次启动 Eclipse 时，需要设置工作空间，在本书中将 Eclipse 安装到 D 盘根目录下，工作空间设置在 D:\eclipse\workspace 中，如图 1-8 所示。

图 1-8 设置工作空间

每次启动 Eclipse 时，都会出现设置工作空间的对话框，如果不需要每次启动都出现该对话框，就可以选中 Use this as the default and do not ask again 复选框，将该对话框屏蔽。

单击 OK 按钮，进入 Eclipse 的欢迎界面，如图 1-9 所示。

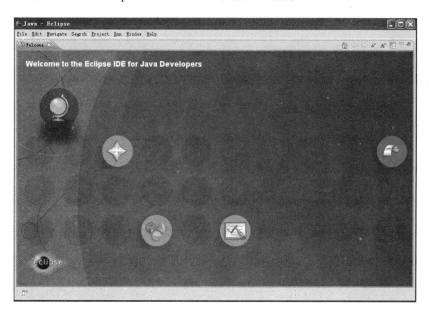

图 1-9 Eclipse 的欢迎界面

Eclipse 工作台是一个 IDE 开发环境，主要由菜单栏、工具栏、资源管理器视图、编辑器、大纲视图、任务视图等组成，如图 1-10 所示。

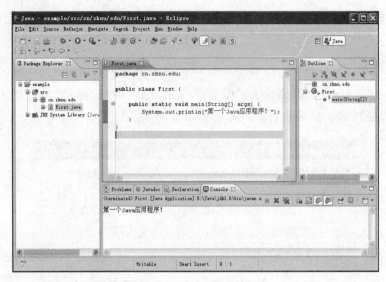

图 1-10 Eclipse 工作台

1.6.2 Eclipse 编写 Java 程序的流程

Eclipse 编写 Java 程序的流程必须经过新建 Java 项目、新建 Java 类、编写 Java 代码和运行程序 4 个步骤，下面分别进行介绍。

1. 新建 Java 项目

在 Eclipse 中选择 File/New/Java Project 菜单命令，如图 1-11 所示。

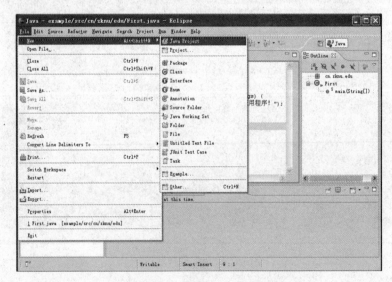

图 1-11 新建 Java 项目

打开 New Java Project（新建 Java 项目），对话框，如图 1-12 所示。

第 1 章　开始 Java 之旅

图 1-12　New Java Project（新建 Java 项目）对话框

单击 Next 按钮，进入 Java 项目构建对话框，在此配置 Java 的构建路径，如图 1-13 所示。

图 1-13　Java 构建设置

在该对话框中，默认 Java 的源文件（Java 文件）放在 src 文件夹，生成的 class 文件放在 bin 文件夹，一般不做修改。单击 Finish 按钮，完成 Java 项目的创建。

完成新建 Java 项目后，在 Package Explorer（包资源管理器）视图中将出现新创建的项目 lesson01，如图 1-14 所示。包资源管理器视图中包含所有已经创建的 Java 项目。

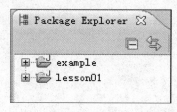

图 1-14 Package Explorer（包资源管理器）

2. 新建 Java 类

在 lesson01 中创建 Java 类。

在 lesson01 上右击鼠标，选择 New/class 菜单命令，弹出 New Java Class（新建 Java 类）对话框，如图 1-15 所示。

图 1-15 New Java Class（新建 Java 类）对话框

各选项含义如下。

- Sourse folder（源文件夹）：用于输入新类的源代码文件夹。
- Package（包）：用于存放新类的包。

- Enclosing type（外出类型）：选择此项，用以选择要在其中封装新类的类型。
- Name（名称）：输入新建 Java 类的名称。
- Modifiers（修饰符）：为新类选择一个或多个访问修饰符。
- Superclass（超类）：为该新类选择超类，默认为 java.lang.Object 类型。
- Interfaces（接口）：编辑新类实现的接口，默认为空。

下面是在新类中选择默认创建哪些方法，分别如下：

- 将 main 方法添加到新类中。
- 从超类复制构造方法到新类中。
- 继承超类或接口的方法，单击 Finish 按钮，即可完成 Java 类的创建。

3. 编写 Java 代码

在 Eclipse 编辑区编写 Java 程序代码，Eclipse 会自动打开源代码编辑器。HelloWorld 类的代码如下：

```
package zknu;
public class HelloWorld {
    public static void main(String[] args) {
        System.out.println("第一个Java应用程序！");
    }
}
```

4. 运行 Java 程序

单击工具栏中 按钮右侧的小箭头，在弹出的下拉菜单中选择 Run As/Java Application 菜单命令，如图 1-16 所示，程序开始运行。程序运行结束后，将在控制台视图中显示程序的运行结果，如图 1-17 所示。

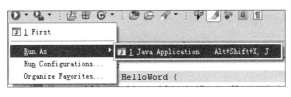
图 1-16 运行 Java 程序

图 1-17 程序运行结果

关键字 class 声明了类的定义，HelloWorld 是描述类名的标识符，整个类的定义包括其所有成员都是在一对大括号（{ }）中完成的，这标志着类定义块的开始和结束。

main() 方法，程序从这里开始执行，所有的 Java 应用程序都必须有一个 main() 方法。main() 方法是所有 Java 应用程序的起始点。

 因为 Java 是区分大小写的，所以 main 与 Main 不同。

关键字 public 是一个访问修饰符，控制类成员的可见度和作用域。

关键字 void 告诉编译器在执行 main() 方法时，不会返回任何值。

关键字 static 允许调用 main() 方法，无须创建类的实例。

String args[] 是传递给 main() 方法的参数，args[] 是 String 类型的数组，String 类型的对象存储字符串。Print() 方法在屏幕上输出以参数方法传递给它的字符串，System 是一个预定义的类，提供对系统类的访问，out 是连接到控制台的输出流。

1.7 要点总结

本章首先介绍了 Java 的特点和目标，然后带领读者完成 Java 开发环境的搭建，其中包括 JDK 的下载与安装、Java 运行环境、JDK 相关环境变量的配置及 JDK 环境的测试方法，通过 Java 程序的运行过程让读者理解 Java 程序的运行原理。最后为了让读者能够快速掌握 Java 语言程序设计的相关语法、技术及其他知识点，介绍了目前流行的 IDE 集成开发工具——Eclipse 及其使用方法，以及编写 Java 程序的流程。

通过对本章内容的学习，读者应该对 Java 语言有初步的认识，并掌握 Java 环境的搭建及开发工具的使用。其中对 Eclipse 开发工具，需要多加练习并从该开发工具自带的教程中了解与掌握更多的知识及使用方法。

1.8 练习题

1. 列举 Java 的三个版本。
2. 简述 Java 程序的开发过程。
3. 简述安装 JDK 需要配置哪些系统变量。
4. 简述 Java 程序的运行原理。
5. 简述 Eclipse 编写 Java 程序的流程。

第 2 章
Java 语言基础

本章重点介绍 Java 程序的基本组成、Java 数据类型、Java 语言的运算符和表达式、Java 语言的流程控制语句，以及组的定义和使用方法、方法和方法的重载。

2.1 Java 程序的基本组成

下面给出一个简单的 Java 程序范例，了解 Java 程序的基本结构。

```java
package zknu;

/**
 * @param TestJavaStructure.java
 * @author chenzhanwei
 * @version v1.0
 */
class Circle{                              // 定义一个圆形类
    final float PI = 3.1415f;              // 声明一个 float 型常量
    int r = 3;                             // 声明一个 int 型变量，初始化值为 3
    /*public float perimeter(int r){       // 求圆周长的方法
            return 2*PI*r;
    }*/
    public float area(int r){              // 求圆面积的方法
            return PI*r*r;
```

```java
    }
}
public class TestJavaStructure {
    public static void main(String[] args) {
        Circle c = new Circle();      // 创建 Circle 的实例化对象
        c.r = 6;                      // 给类的成员变量 r 赋值
//      System.out.println(«圆的周长为: » + c.perimeter(c.r));
        System.out.println(«圆的面积为: » + c.area(c.r));
    }
}
```

程序运行结果如下：

```
圆的面积为：113.093994
```

程序的注释不仅有助于提高程序的可读性，还可以屏蔽一些暂时不用的语句，等需要时直接取消此语言的注释即可。在 Java 中，根据功能的不同可分为单行注释、多行注释（块注释）和文档注释三种。

（1）单行注释

在注释内容前面加"//"，Java 编译器会忽略这部分信息，如程序中下面的语句：

```
final float PI = 3.1415f;                    // 声明一个 float 型常量
```

（2）多行注释

在注释内容前面加"/*"，在注释内容后面加"*/"，一般注释内容为多行，如程序中对圆周长方法的注释就是多行注释。

（3）文档注释

程序中"/** 注释内容 */"形式为文档注释，这种方法注释的内容会被解释成程序的正式文档，并能包含在如 javadoc 之类工具生成的文档中，用以说明该程序。

2. class 和 public class

在 Java 中声明一个类的方式主要有两种，即 class 类名称和 public class 类名称。

类是 Java 的基本存储单元，Java 中所有的操作都是由类组成的。一般习惯把 main 方法放在 public class 声明的类中，public static void main(String[] args) 是程序的主方法，即所有的程序都以此方法为起点并运行下去。public class 类名称的"类名称"必须与文件名相同。

在一个 Java 文件中可以有多个 class 类的定义，但只能有一个 public class 的定义。

3. 标识符和关键字

Java 语言中的类名、接口名、对象名、方法名、常量名、变量名等通称为标识符，由程序员自己定义。Java 语言中，标识符是以字母、下画线（_）和美元符（$）开始的一个字符序列，后面可以跟字母、下画线、美元符和数字。建议最好以字母开头，尽量不要包含其他符合，如程序中的类名"Circle""TestJavaStructure"等。

　　Java 语言还定义了一些具有专门意义和用途的关键字，也称保留字。Java 中的关键字全部用小写字母表示，它们不能被当作合法的标识符使用。Java 中的关键字有 abstract、break、byte、boolean、catch、case、class、char、continue、default、double、do、else、extends、false、final、float、for、finally、if、import、implements、int、interface、instanceof、long、length、native、new、null、package、private、protected、public、return、switch、synchronized、short、static、super、try、true、this、throw、throws、threadsafe、transient、void 和 while。

　　这些关键字不需要读者去强记，如果开发中使用了这些关键字，编译器就会自动提示错误。另外，true、false、null 虽然不是关键字，但是作为一个单独的标识类型也不能直接使用。

　　在定义标识符时，尽量遵循"见其名知其意"的原则。Java 标识符的具体命名规则如表 2-1 所示。

表2-1 Java标识符的命名规则

元素	规范	示例
类名、接口名	首字母大写	Person Student SystemManager
变量名、数组名	Camel规则，小写开头	ageValue salary printInformation
函数名（方法名）	Camel规则，小写开头	setCourse getAge setUserName
包名	全部小写	com.zknu.czw sam.gover
常量名	全部大写	MAX_VALUE

4．常量和变量

　　所谓常量，就是值永远不允许被改变的量。如果要声明一个常量，就必须用关键字 final 修饰，如程序中常量的声明和赋值。

```
final float PI = 3.1415f;        // 声明一个 float 型常量
```

　　在声明常量时，必须立即为其赋值，即初始化。

　　所谓变量，就是值可以改变的量。变量利用声明的方式将内存中的某个内存块保留下来以供程序使用。变量可以用来存放数据，而使用之前必须先声明它的数据类型，也可以在声明时立即为其赋值，如程序中变量的声明和初始化。

```
int r = 3;                       // 声明一个 int 型变量，初始化值为 3
```

　　变量的作用域是一个程序的区域。变量声明时就决定了变量的作用域。在一个确定的域中，变量名应该是唯一的。

　　变量的作用域可以是成员变量作用域、本地变量作用域、方法参数作用域或异常处理参数作用域。

作用域可以嵌套。外部作用域的变量对于内部作用域是可见的，但内部作用域的变量对外部作用域是不可见的。

 虽然内部作用域的变量对外部作用域是不可见的，但建议开发者避免内外作用域使用相同的变量名。

2.2 Java 语言的数据类型

Java 语言的数据类型分为原始类型（简单类型）和引用类型（复合类型）。

原始数据类型包括以下 8 种。

（1）整数类型：byte、short、int 和 long。

（2）浮点类型：float 和 double。

（3）字符类型：char。

（4）布尔类型：boolean。

引用数据类型包括类、接口和数组三种。

2.2.1 整数类型

Java 定义了 4 个整数类型，即 byte、short、int 和 long，它们都是带符号的。

1. byte

byte 即字节型，是最小的整数类型，所占位数为 8 位。取值范围为 $-2^7 \sim 2^7-1$，即 -128 ~ 127，常用于数据流的处理。

2. short

short 即短整型，所占位数为 16 位。取值范围为 $-2^{15} \sim 2^{15}-1$，即 -32768~32767，主要用于 16 位计算机，现在很少使用。

3. int

int 即整型，所占位数为 32 位。取值范围为 $-2^{31} \sim 2^{31}-1$，即 -2147483648~2147483647。整型是常用的数据类型之一，经常用于循环的计数器和数组的下标。

4. long

long 即长整型，所占位数为 64 位。取值范围为 $-2^{63} \sim 2^{63}-1$，即 -9223372036854775808~9223372036854775807。长整型也是常用的数据类型之一，用来表示超过整型的数字比，如时间的毫秒数等。

2.2.2 浮点类型

1. float

float 即单精度浮点型，所占位数为 32 位。取值范围为 1.4E-45 ～ 3.4028235E38，常用于对小数位精度要求不是很高的数字。

2. double

double 即双精度浮点型，所占位数为 64 位。取值范围为 4.9E-324~1.7976931348623157E308，常用于需要计算精确度要求很高的情况。

2.2.3 字符类型

char 即字符型，Java 使用 Unicode 码代表字符，这一点决定了 Java 中 char 所占位数不同于 C/C++ 的 8 位而是 16 位。因为 char 是无符号的，所以取值范围为 0~65535。

2.2.4 布尔类型

boolean 即布尔类型，只包含 True 和 False 两个值，多用于流程控制语句的条件表达式。

2.2.5 基本数据类型的默认值

在 Java 中，若在变量声明时没有给变量赋初值，则会给该变量赋默认值。表 2-2 列出了基本数据类型的默认值。

表2-2 基本数据类型的默认值

序号	数据类型	默认值
1	Byte	(byte)0
2	short	(short)0
3	Int	0
4	Long	0L
5	Float	0.0f
6	double	0.0d
7	Char	\u0000(空)
8	boolean	false

2.2.6 类型转换

由于 Java 数据类型在定义时就已经确定了，因此不能随意转换为其他的数据类型，但 Java 允许用户有限度地做类型转换处理。所有基本数据类型中，只有 boolean 不能与其他数据类型相互转换，剩下的数据类型可以相互转换，其中 byte 数据类型级别最低，double 数据类型级别最高。转换时根据转换方向的不同，可分为"自动转换"和"强制转换"。

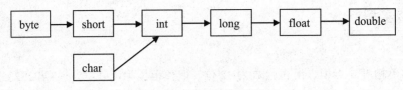

图 2-1 基本数据类型转换规则

1. 自动转换

沿着图 2-1 的箭头方向进行转换，不损失精度的转换称为自动转换，也称为隐式转换。即满足以下两个条件，数据类型之间的转换是自动进行的。

（1）进行转换的两种类型是兼容的。

（2）目标类型的范围大于源类型的范围。

2. 强制转换

沿着箭头相反的方向进行转换，有可能损失精度的转换称为强制转换，也称为显式转换。即不满足类型自动转换的条件下仍然希望进行类型转换，就只能进行强制转换。通过在源类型的变量或数值前加（目标类型）进行强制转换，这也是强制转换也称为显式转换的原因。

下面程序演示了自动转换和强制转换两种情况：

```java
public class Convert {
    public static void main(String[] args) {
        byte a = 10;                // 定义 byte 类型的变量
        int b = a;                  // byte 自动转换为 int
        long c = b;                 // int 自动转换为 long
        double d = b;               // int 自动转换为 double
        float f = 3.14f;            // 定义 float 类型的变量
        double e = f;               // float 自动转换为 double
        int g = (int)f;             // float 强制类型转换为 int
        int h = (int)e;             // double 强制类型转换为 int
    }
}
```

> **注意** 开发者应尽量避免无谓的强制类型转换。

对于类型转换还存在一个容易被忽视的错误。先来看一个例子：

```java
public class ConvertDemo {
    public static void main(String[] args) {
        byte b = 10;                           // 定义 byte 类型的变量
        byte result = (byte) (b + 4);          // int 强制转换为 byte
        System.out.println("<--result 为 "+result+"-->");
```

```
    }
}
```

这段程序代码的运行结果如下:

```
<--result 为 14-->
```

细心的读者可能会发现,上面的例子中进行了一次强制转换。这次强制转换是否有必要呢?答案是肯定的。因为在 Java 的表达式中会进行类型提升,这种表达式中的类型提升是自动进行的。提升的规则就是将表达式运算结果的类型提升为所有操作数数据类型中范围最大的数据类型。在上面的例子中常数 4 被认为是 int 类型,根据这个规则,也就不难理解为什么需要进行强制类型转换了。

 不要忽视常数的默认类型对运算结果的影响。

2.3 运算符和表达式

参与运算的常量、变量和表达式统称为操作数。连接操作数完成运算的符号称为运算符。表达式是由操作数和运算符按一定的语法形式组成的符号序列。在高级程序设计语言中,运算符通常分为算术运算符、位运算符、关系运算符和逻辑运算符 4 类。而从操作数的个数又可分为一元操作符、二元操作符和三元操作符。

2.3.1 赋值运算符

在介绍算术运算符、位运算符、关系运算符和逻辑运算符之前,先简单说明一下赋值运算符。赋值运算符用"="表示,作用就是给变量赋值。赋值运算符比较容易理解并且前面的例子也都使用过,这里就不再赘述了。

2.3.2 算术运算符

算术运算符,顾名思义用于在数学表达式中进行算术运算。算术运算符可以用于除布尔类型以外的所有原始数据类型。

算术运算符包括基本算术运算符、简写算术运算符和递增递减运算符。

基本算术运算符如下:

(1)"+"代表加法,二元操作符。

(2)"-"对于二元操作数代表减法,对于一元操作符代表取负。

(3)"*"代表乘法,二元操作符。

（4）"/"代表除法，二元操作符。

（5）"%"代表求模，二元操作符。

如果用 op 代表上述基本运算符，var1 和 var2 就分别代表两个操作数。形式如下：

```
var1 = var1 op var2;
```

可以简写为：

```
var1 op= var2;
```

对于这种情况，上述 5 个基本算术运算都有其简写的算术运算符，分别为 "+=" "-=" "*=" "/=" 和 "%="。

递增递减运算符 "++" 和 "--" 的用法都有两种。如果用 var1 和 var2 表示两个变量，那么这两种用法可以表示为：

```
var1++、var1-- 和 ++var1、--var1
```

对于上述语句，这两种用法没有区别，都分别表示在 var1 基础上加 1 和减 1。而对于下面的语句，这两种用法是有区别的，以 "++" 运算符为例，"--" 运算符同理。

```
var2 = var1++;
等价于
var2 = var1;
var1 = var+1;
而
var2=++var1;
等价于
var1 = var+1;
var2 = var1;
```

 对于 byte 类型、short 类型和 char 类型的变量，上面的等价关系并不成立。对于这三种类型的变量，执行 var1 = var+1;语句会产生错误，因为常数 1 默认为 int 类型，根据前面提到的表达式中类型提升原则，这样的语句相当于将 int 类型隐式地转换为范围低于它的类型。而这样做是不可能通过编译的，必须进行类型的强制转换才行。

2.3.3 位运算符

位运算符用于对整数类型操作数的二进制编码的位运算。位运算符可以用于整数类型和字符类型。位运算符如下：

(1) "~"代表按位非，一元运算符。

(2) "&"代表按位与，二元操作符。

(3) "|"代表按位或，二元操作符。

(4) "^"代表按位异或，二元操作符。

(5) ">>"代表右移，二元操作符。

(6) "<<"代表左移，二元操作符。

(7) ">>>"代表右移且空出的位用 0 填充，二元操作符。

(8) "&="代表"&"运算符的简写运算符，其用法与算术运算符相同，二元操作符。

(9) "|="代表"|"运算符的简写运算符，其用法与算术运算符相同，二元操作符。

(10) "^="代表"^"运算符的简写运算符，其用法与算术运算符相同，二元操作符。

(11) ">>="代表">>"运算符的简写运算符，其用法与算术运算符相同，二元操作符。

(12) "<<="代表"<<"运算符的简写运算符，其用法与算术运算符相同，二元操作符。

(13) ">>>="代表">>>"运算符的简写运算符，其用法与算术运算符相同，二元操作符。

2.3.4 关系运算符

关系运算符用于判断操作数之间的关系，运算的结果是布尔类型的，即 True 或 False。关系运算符如下：

(1) "=="代表等于关系，二元操作符。

(2) "!="代表不等于关系，二元操作符。

(3) ">"代表大于关系，二元操作符。

(4) "<"代表小于关系，二元操作符。

(5) ">="代表不小于关系，二元操作符。

(6) "<="代表不大于关系，二元操作符。

 初学者容易混淆赋值运算符"="和关系运算符"=="。

"=="和"!="可以用于所有数据类型。其他的关系运算符可以用于除布尔类型之外的所有原始数据类型。

2.3.5 三元运算符

"?:"是 Java 中唯一的一个三元运算符。如果 var 表示类型为布尔类型的操作数，var1 和 var2 表示类型相同的操作数，那么这个三元运算符的用法如下：

```
var?var1:var2
```

如果 var 的值为 True，那么表达式结果为 var1 的值；否则为 var2 的值。

2.3.6 运算符优先级

多个运算符参与运算时会从左到右按照运算符优先级别的高低依次进行运算。各运算符的优先级如表 2-3 所示（1 代表最高优先级、15 代表最低优先级）。

表2-3 各运算符的优先级

优先级	运算符
1	[ ]、( )
2	++、--、!、~
3	*、/、%
4	+、-
5	>>、>>>、<<
6	>、<、>=、<=
7	==、!=
8	&
9	^
10	|
11	&&
12	||
13	?:
14	=、+=、-=\、*=、/=、%=、^=
15	&=、|=、<<=、>>=、>>>=

虽然运算符有优先级，但还是建议用括号控制运算顺序。用括号控制运算顺序的代码可读性强。

2.4 流程控制语句

程序通过流程控制语句完成对语句执行顺序的控制，如循环执行、选择执行等。Java 中的流程控制语句与 C++ 的流程控制语句并无太大差异。流程控制语句可分为选择、循环和跳转三大类。

2.4.1 选择语句

选择语句的作用是根据判断条件选择执行不同的程序代码。选择语句包括 if-else 语句和 switch-case 语句。

1. if-else 语句

```
if-else 语句的第一种形式:
```

```
if(布尔表达式){
程序代码块 1;
} else {
程序代码块 2;
}
```

其中，else 块是可选的。

if-else 语句的执行过程：如果布尔表达式为 true，就运行程序代码块 1；否则运行程序代码块 2。

【示例 2-1】

if-else 语句具体的使用方法可以参看下面的例子。

```
public class IfElseDemo {
    /**
     * 根据输入的合法成绩判断是否合格
     * @param result
     */
    private static void judge(int result) {
        System.out.println("<-- 成绩为 " + result + "-->");
        if (result >= 60) {
            System.out.println("<-- 恭喜，这个成绩合格！-->");
        } else {
            System.out.println("<-- 很遗憾，这个成绩不合格！-->");
        }
    }
    public static void main(String[] args) {
        int firstResult = 80;            // 定义 int 类型变量
        int secondResult = 45;           // 定义 int 类型变量
        judge(firstResult);
        judge(secondResult);
    }
}
```

这段程序代码的运行结果如下：

```
<-- 成绩为 80-->
<-- 恭喜，这个成绩合格！-->
<-- 成绩为 45-->
<-- 很遗憾，这个成绩不合格！-->
```

【示例 2-2】

if-else 语句是可以嵌套使用的。例如：

```java
public class NestIfElseDemo {
    /**
     * 根据成绩判断是否合格
     * @param result
     */
    private static void judge(int result) {
        System.out.println("<-- 成绩为 " + result + "-->");
        if (result < 0 || result > 100) {
System.out.println("<-- 对于百分制这个成绩不合法,请检查输入的成绩! -->");
        } else {
            if (result >= 60) {
                System.out.println("<-- 恭喜,这个成绩合格! -->");
            } else {
                System.out.println("<-- 很遗憾,这个成绩不合格! -->");
            }
        }
    }
    public static void main(String[] args) {
        int firstResult = 80;            // 定义 int 类型变量
        int secondResult = 45;           // 定义 int 类型变量
        int thirdResult = -10;           // 定义 int 类型变量
        judge(firstResult);
        judge(secondResult);
        judge(thirdResult);
    }
}
```

这段程序代码的运行结果如下:

```
<-- 成绩为 80-->
<-- 恭喜,这个成绩合格! -->
<-- 成绩为 45-->
<-- 很遗憾,这个成绩不合格! -->
<-- 成绩为 -10-->
<-- 对于百分制这个成绩不合法,请检查输入的成绩! -->
```

if-else 语句的第二种形式如下:

```
if( 布尔表达式 1){
```

程序代码块 1;

```
} else if( 布尔表达式 2) {
```

程序代码块 2;

```
    } else if(布尔表达式 3) {
```

程序代码块 3;

```
    } else if(布尔表达式 n) {
```

程序代码块 n;

```
    } else {
```

程序代码块;

```
}
```

其中，else 块是可选的。

程序依次判断布尔表达式，如果判断为 true，就执行与之对应的程序代码块，而后面的布尔表达式全部忽略；如果所有的布尔表达式都为 false，就执行 else 对应的代码块。这种形式可以等价替换上面介绍的嵌套 if-else 语句，例如：

```java
public class IfElseIfDemo {
    /**
     * 根据成绩判断是否合格
     * @param result
     */
    private static void judge(int result) {
        System.out.println("<-- 成绩为 " + result + "-->");
        if (result < 0 || result > 100) {
System.out.println("<-- 对于百分制这个成绩不合法，请检查输入的成绩！-->");
        } else if (result >= 60) {
            System.out.println("<-- 恭喜，这个成绩合格！-->");
        } else {
            System.out.println("<-- 很遗憾，这个成绩不合格！-->");
        }
    }
    public static void main(String[] args) {
        int firstResult = 80;           // 定义 int 类型变量
        int secondResult = 45;          // 定义 int 类型变量
        int thirdResult = -10;          // 定义 int 类型变量
        judge(firstResult);
        judge(secondResult);
        judge(thirdResult);
    }
}
```

这段程序代码的运行结果如下：

```
<-- 成绩为 80-->
<-- 恭喜，这个成绩合格！ -->
<-- 成绩为 45-->
<-- 很遗憾，这个成绩不合格！ -->
<-- 成绩为 -10-->
<-- 对于百分制这个成绩不合法，请检查输入的成绩！ -->
```

2. switch-case 语句

switch-case 语句的形式如下：

```
switch（表达式）{
case 选择值1 ：   程序代码块1；
break；
case 选择值2 ：   程序代码块2；
break；
············
case 选择值n ：   程序代码块n；
break；
default：        程序代码块；
}
```

switch 表达式可以是 byte、short、char 和 int 类型中的一种。case 的值必须是与 switch 表达式类型一致的常量并且不能重复。

switch 语句的执行过程：switch 表达式的值与 case 的常量依次比较，如果相等，就执行相应 case 后面的所有代码；如果没有与 switch 表达式的值相等的常量，就执行 default 后面的代码。

switch 语句具体的使用方法可以参看下面的例子。

```java
public class SwitchDemo {
    public static void main(String[] args) {
        int x = 3;                              // 声明整型变量 x
        int y = 6;                              // 声明整型变量 y
        char oper = '+';                        // 声明字符变量 ch
        switch (oper) {                         // 将字符作为 switch 的判断条件
            case '+':{                          // 判断字符内容是否是 "+"
                System.out.println("x+y=" + (x+y));
                break;                          // 退出 switch
            }
            case '-':{                          // 判断字符内容是否是 "-"
                System.out.println («x-y=» + (x-y));
                break;                          // 退出 switch
            }
            case '*':{                          // 判断字符内容是否是 »*»
```

```
                    System.out.println("x*y=" + (x*y));
                    break;                  // 退出 switch
                }
                case '/':{                  // 判断字符内容是否是"/"
                    System.out.println("x/y=" + (x/y));
                    break;                  // 退出 switch
                }
                default:{                   // 其他字符
System.out.println("未知的操作!");
break;                                      // 退出 switch
                }
            }
        }
}
```

这段程序代码的运行结果如下：

x+y=9

读者可以自行将 oper 中的操作修改为 "+" "-" "*" "/" 等。如果设置的是一个未知的操作，那么程序将提示"未知的操作！"。

 如果每个 case 的程序代码块的最后没有 break 语句，那么程序将会执行程序代码块后面的所有代码，读者可自行测试。

2.4.2 循环语句

循环语句的作用是反复执行一段代码，直到不能满足循环条件为止。循环语句包括 for 语句、while 语句和 do-while 语句。

1. for 语句

for 语句的形式如下：

```
for ( 初始化 ; 循环条件 ; 迭代部分 ){
程序代码块；
}
```

从理论上讲，初始化、循环条件和迭代部分都是可选的。如果程序代码块中没有跳转语句，那么下面的形式将会是一个无限循环，也称作死循环。

```
for(; ;){
程序代码块；
}
```

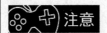 死循环不一定是错误的。事实上,在 Java 线程中经常会主动构造死循环。

for 语句的执行过程:首先执行初始化的代码,这部分代码只执行一次,然后判断是否满足循环条件,循环条件是布尔表达式。如果满足,就执行循环体中的程序代码,最后执行迭代部分。再判断是否满足循环条件,如此循环往复直到不满足循环条件为止。如果不满足,就执行 for 语句后面的程序代码。

【示例 2-3】

for 语句具体的使用方法可以参看下面的例子。

```
public class ForDemo {
    public static void main(String[] args) {
        int sum = 0;                              // 定义变量保存累加结果
        for (int i = 1; i <=10; i++) {
            sum += i;                             // 执行累加操作
        }
        System.out.println("1-->10 累加结果为: " + sum);    // 输出累加结果
    }
}
```

这段程序代码的运行结果如下:

```
1-->10 累加结果为: 55
```

【示例 2-4】

for 语句是可以嵌套的。嵌套的循环语句就是通常所说的多重循环。下面的例子就是一个二重循环。

```
public class ForNestedDemo {
    public static void main(String[] args) {
        for (int i = 1; i < 9; i++) {                      // 第一层循环
            for (int j = 1; j <= i; j++) {                 // 第二层循环
                System.out.print(i + "*" + j + "=" + (i*j) + "\t");
            }
            System.out.print("\n");                        // 换行
        }
    }
}
```

这段程序代码的运行结果如下:

```
1*1=1
2*1=2    2*2=4
```

```
3*1=3      3*2=6  3*3=9
4*1=4      4*2=8  4*3=12 4*4=16
5*1=5      5*2=10 5*3=15 5*4=20 5*5=25
6*1=6      6*2=12 6*3=18 6*4=24 6*5=30 6*6=36
7*1=7      7*2=14 7*3=21 7*4=28 7*5=35 7*6=42 7*7=49
8*1=8      8*2=16 8*3=24 8*4=32 8*5=40 8*6=48 8*7=56 8*8=64
9*1=9      9*2=18 9*3=27 9*4=36 9*5=45 9*6=54 9*7=63 9*8=72 9*9=81
```

JDK1.5 后为了方便数组的输出，提供了 foreach 语法，格式如下：

```
for each(数据类型 变量名称 : 数组名称){
    语句序列；
}
```

使用 foreach 语法输出数组内容。例如：

```
public class ArrayDemo {
    public static void main(String[] args) {
        int[] score = {60,89,86,90,73,56};
        for each(int i : score) {
            System.out.print(i + "\t");
        }
    }
}
```

程序运行结果如下：

```
60 89      86      90      73      56
```

 虽然 Java 中提供了 foreach 语法，但是从实际的应用来看，还是使用最原始的输出操作比较合适，所以不建议读者过多地使用 foreach 语法输出。

2. while 语句

while 语句的形式如下：

```
while（循环条件）{
程序代码块；
}
```

while 语句比较简单，循环条件也必须是布尔表达式。

while 语句的执行过程：首先判断是否满足循环条件。如果满足，就执行循环体中的程序代码；如果不满足，就跳过循环体执行 while 语句后面的程序代码。例如：

```
public class WhileDemo {
    public static void main(String[] args) {
```

```
        int x = 1;                          // 定义整型变量
        int sum = 0;                        // 定义整型变量保存累加结果
        while (x <= 10) {                   // 判断循环结果
            sum += x;                       // 执行累加结果
            x++;                            // 修改循环条件
        }
        System.out.println("1-->10 累加结果为：" + sum);  // 输出累加结果
    }
}
```

这段程序代码的运行结果如下：

```
1-->10 累加结果为：55
```

与 for 循环语句一样，while 语句也可以嵌套。由于 while 语句比较简单，这里就不再举例了。

3. do-while 语句

do-while 语句的形式如下：

```
do {
程序代码块；
} while (循环条件);
```

do-while 语句与 while 语句很相似，区别在于：do-while 语句是先执行循环体内的程序代码块，然后判断是否满足循环条件。下面的例子说明了 do-while 语句的使用方法。

```
public class DoWhileDemo {
    public static void main(String[] args) {
        int x = 1;                          // 定义整型变量
        int sum = 0;                        // 定义整型变量保存累加结果
        do {                                // 判断循环结果
            sum += x;                       // 执行累加结果
            x++;                            // 修改循环条件
        } while (x <= 10);                  // 判断循环
        System.out.println("1-->10 累加结果为：" + sum);  // 输出累加结果
    }
}
```

这段程序代码的运行结果如下：

```
1-->10 累加结果为：55
```

4. 循环中的中断

在 Java 语言中，可以使用如 break、continue 等中断语句。站在结构化程序设计的角度上，并不鼓励开发者使用中断语句。

(1) break 语句

break 语句可以强迫程序中断循环。当程序执行到 break 语句时,即会离开循环,继续执行循环外的下一个语句。如果 break 语句出现在嵌套循环中的内层循环,就会跳出当前层的循环。以下面的 for 循环为例,在循环主体中有 break 语句时,程序执行到 break,即会离开循环主体,继续执行循环外层的语句。

```java
public class BreakDemo {
    public static void main(String[] args) {
        for (int i = 0; i < 10; i++) {          // 使用for循环
            if (i == 3) {                        // 如果i的值为3,就退出整个循环
                break;                           // 退出整个循环
            }
            System.out.print("i=" + i + " ");   // 打印信息
        }
    }
}
```

这段程序代码的运行结果如下:

```
i=0   i=1   i=2
```

从程序的运行结果可以发现,当 i 的值为 3 时,判断语句满足,就执行 break 语句退出整个循环。

(2) continue 语句

continue 语句可以强迫程序跳到循环的起始处。当程序运行到 continue 语句时,会停止运行剩余的循环主体,回到循环的开始处继续执行。下面的例子说明了 continue 语句的使用方法。

```java
public class ContinueDemo {
    public static void main(String[] args) {
        for (int i = 0; i < 10; i++) {          // 使用for循环
            if (i == 3) {
                continue;                        // 退出一次循环
            }
            System.out.print("i=" + i + " ");   // 打印信息
        }
    }
}
```

这段程序代码的运行结果如下:

```
i=0   i=1   i=2   i=4   i=5   i=6   i=7   i=8   i=9
```

从程序的运行结果中可以发现，当 i 的值为 3 时，程序并没有向下执行输出语句，而是退回到了循环判断处继续向下执行，也就是说，continue 只是中断了一次的循环操作。

2.5 数组与方法

数组属于引用数据类型。在数组中有以下几个概念：

- 数组的名字：数组有一个名字；
- 数组的类型：数组中所有的数据具有相同的类型；
- 数组的元素：数组中的一个数据称为一个元素；
- 数组的索引：元素的序号，第一个元素索引从 0 开始；
- 数组的长度：整个数组的元素个数。

数组是一组相关数据的集合，一个数字实际上就是一连串的变量。数组按照使用可以分为一维数组、二维数组和多维数组。

2.5.1 一维数组

一维数组就是一组具有相同数据类型的有序变量集合。使用数组之前要先声明，方法如下：

```
数据类型 数组名[ ] 或者 数据类型 [ ] 数组名；
```

例如，声明一个整型的数组 buffer：

```
int[] buffer;
```

由于数组类型是引用类型（复合类型），因此声明的数组变量是引用。因为还没有为数组变量分配相应的内存空间，所以为空，这也是为什么声明数组时并不要求指明数组大小的原因。为一维数组分配内存空间的方法如下：

```
数组名 = new 数据类型 [数组大小];
buffer = new int[5];
数据类型 数组名[ ] = new 数据类型 [数组大小];
```

或者

```
数据类型 [ ] 数组名 = new 数据类型 [数组大小];
int[]buffer = new int[50]    //声明数组的同时创建数组
```

或者

```
int[]buffer = new int[]{10,20,30,40,60};    // 因为格式比较烦琐，所以很少使用
```

使用关键字 new 创建数组时所有元素已经被初始化，即元素都是默认值，这种初始化称为"动态初始化"。

还有一种不使用关键字 new，在声明数组的同时就完成了创建和初始化工作，称为"静态初始化"。

```
int[]  buffer = {2,3,4,1,9}   // 不使用 new，必须写在一行。
```

分配了内存空间的数组就可以通过声明的数组名和下标来访问数组中的元素了。下标从 0 开始到数组大小 -1。不同于 C 和 C++，Java 中会进行数组越界检查，也就是说，使用数组名和超过数组大小 -1 的下标进行访问是被禁止的。

下面先对 Java 的内存管理进行介绍，然后通过一个例题分析数组的内存操作过程。

本书把 Java 的内存分为代码区、数据区、栈内存和堆内存 4 个区。

- 代码区（code segment）：主要存放程序代码和对象的方法，并且是多个对象共享一块存储区。
- 数据区（data segment）：存放静态（static）变量和字符串变量。
- 栈内存（stack）：存放对象引用、局部变量、基础数据类型、方法的形参、方法的引用参数等。在使用完毕或生命周期完成后就直接回收，不需要垃圾回收机制。
- 堆内存（heap）：在运行时以随机顺序进行存储空间分配和收回的内存管理模型。Java 对象的内存总是在 heap 中分配，需要垃圾回收机制。

下面的例子是对数组的声明、创建、赋值和输出操作，以及数组长度的取得方法。

```java
public class ArrayDemo01 {
  public static void main(String[] args) {
        int[] score = null;                 // 声明数组，但未开辟堆内存空间
        score = new int[3];                 // 为数组开辟堆内存空间
        for (int i = 0; i < score.length; i++) {
// 输出数组的全部内容
System.out.println("score["+i+"]=" + score[i]);
        }
        for (int i = 0; i < score.length; i++) {
score[i] = 2*i;                             // 为每一个元素赋值
        }
        for (int i = 0; i < score.length; i++) {
// 输出数组的全部内容
System.out.println("score["+i+"]=" + score[i]);
        }
    }
```

}
```

这段程序代码的运行结果如下：

```
score[0]=0
score[1]=0
score[2]=0
score[0]=0
score[1]=2
score[2]=4
```

执行的内存操作流程如图 2-2 所示。

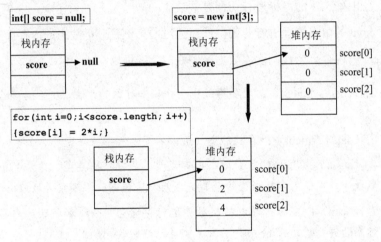

图 2-2 内存操作流程

从程序运行结果和内存操作流程可以看到，在栈内存中保存的永远是数组的名称，只开辟了栈内存空间的数组是永远无法使用的，必须有指向的堆内存才可以使用。要想开辟新的堆内存，必须使用关键字 new，然后将此堆内存的使用权交给对应的栈内存空间，并且一个堆内存可以同时被多个栈内存空间所指向。如下面数组的赋值，如图 2-3 所示。

```
int[] a = {1,2,3};           // 声明数组 a 并静态初始化
  int[] b = a;               // 把数组 a 赋值给数组 b
```

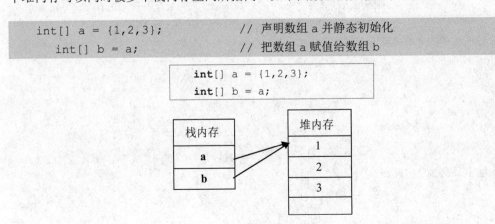

图 2-3 数组赋值
```

 **提示** 因为数组是引用数据类型，所以后面学到的类、接口等引用数据类型的内存分配操作与数组相同。

2.5.2 二维数组

如果把一维数组看作是一句话，二维数组就可以看作是一个表。二维数组的声明方法与一维数组类似，内存的分配也要使用关键字 new 来完成。其声明的格式如下：

数据类型 数组名 [ ][ ] 或者 数据类型 [ ][ ] 数组名；

二维数组分配内存的方法如下：

数组名 = new 数据类型 [行数组大小][列数组大小]；

与一维数组不同的是，二维数组在分配内存时，必须告诉编译器二维数组行与列的个数。例如：

int[][] a = new int[4][3];   // 声明整型数组 a，同时为其开辟一块内存空间

二维数组 a 占用的内存空间为多少字节？

二维数组的定义及应用如下：

```
int score1[][] = new int[4][3] ;            // 声明并实例化二维数组
    score1[0][1] = 30 ;                     // 为数组中的内容赋值
    score1[1][0] = 31 ;                     // 为数组中的内容赋值
    score1[2][2] = 32 ;                     // 为数组中的内容赋值
    score1[3][1] = 33 ;                     // 为数组中的内容赋值
    score1[1][1] = 30 ;                     // 为数组中的内容赋值
    for(int i=0;i<score1.length;i++){
        for(int j=0;j<score1[i].length;j++){
            System.out.print(score1[i][j] + "\t") ;
        }
        System.out.println("") ;
    }
```

读者可自行编写，并运行其结果。

还有一种不常用的为二维数组分配内存的方法。

数组名 = new 数据类型 [行数组大小][ ];
数组名 [0] = new 数据类型 [数组大小];
数组名 [1] = new 数据类型 [数组大小];
… … …
… … …
… … …

零基础轻松学 Java

```
数组名 [ 行数组大小 -1] = new 数组类型 [ 数组大小 ];
```

这种分配内存的方式也不难理解，特殊的地方在于每一次分配列数组大小时其大小可以不同，这样就可以构造出具有固定行数，而列数却不固定的不规则数组。

下面的例子演示了二维的不规则数组。

```java
public class MultiArrayDemo {
    public static void main(String[] args) {
        int[][] firstArray;            // 声明第一个 int 类型二维数组
        // 用另一种方法声明第二个 int 类型二维数组，同时分配内存空间
        int secondArray[][] = new int[2][2];
        // 用另一种方法给第一个 int 类型二维数组分配内存空间
        firstArray = new int[3][];
        firstArray[0] = new int[1];
        firstArray[1] = new int[2];
        firstArray[2] = new int[3];
        // 访问第一个 int 数组顺序填入自然数
        firstArray[0][0] = 1;
        firstArray[1][0] = 2;
        firstArray[1][1] = 3;
        firstArray[2][0] = 4;
        firstArray[2][1] = 5;
        firstArray[2][2] = 6;
        // 访问第二个 int 数组顺序填入自然数
        secondArray[0][0] = 1;
        secondArray[0][1] = 2;
        secondArray[1][0] = 3;
        secondArray[1][1] = 4;
    }
}
```

二维数组也可以在声明时就被初始化，方法类似于一维数组。例如：

```java
public class MultiArrayInit {
    public static void main(String[] args) {
        char[][] firstArray = {{'1', '2' }, {'3', '4' } };
        char secondArray[][] = {{ '1' }, {'2', '3' }, {'4', '5', '6' } };
        System.out.println("<-- 第一个二维数组开始 -->");
        System.out.println(firstArray[0]);
        System.out.println(firstArray[1]);
        System.out.println("<-- 第一个二维数组结束 -->");
        System.out.println("<-- 第二个二维数组开始 -->");
        System.out.println(secondArray[0]);
        System.out.println(secondArray[1]);
```

```
            System.out.println(secondArray[2]);
            System.out.println("<-- 第二个二维数组结束 -->");
    }
}
```

多维数组很少使用,本书不再讲述。

2.5.3 方法

方法就是一段可重复使用的代码段,可以简化代码的编写量。有些书中把方法称为函数,两者本身并没有区别,属于同样的概念,只是称呼不同而已。

1. 方法的定义和调用

Java 中可以使用多种方式定义方法,如前面常用的 main 方法,在声明处加上了 public static 关键字,static 关键字将在后面的章节详细讲解。方法的定义格式如下:

```
[ 修饰限定符 ]  返回值类型  方法名称(类型  参数1,类型  参数2,…){
    语句序列;
    [return  表达式];
}
```

如果方法没有返回值,就要在"返回值类型"明确写出 void,此时方法中的 return 语句可以省略。方法执行完毕后,无论是否存在返回值都要返回到方法的调用处向下执行。方法名称要遵循 Java 标识符的命名规则。参数列表可以为空,也可以有多个。

下面的例子演示了方法的定义和调用。

```
public class MethodDemo01 {
  public static void main(String[] args) {
        print();
        int one = addMethod1(10,20) ;              // 调用整型的加法操作
        float two = addMethod2(10.3f,13.3f) ;      // 调用浮点数的加法操作
        System.out.println("addMethod1 的计算结果: " + one) ;
        System.out.println("addMethod2 的计算结果: " + two) ;
  }
  public static void print(){
        System.out.println(" 演示方法的调用! ");
  }
  // 定义方法,完成两个数字的相加操作,方法返回一个 int 型数据
  public static int addMethod1(int x,int y){
        int temp = 0 ;                  // 方法中的参数,是局部变量
        temp = x + y ;                  // 执行加法计算
        return temp ;                   // 返回计算结果
  }
  // 定义方法,完成两个数字的相加操作,方法的返回值是一个 float 型数据
```

```
    public static float addMethod2(float x,float y){
        float temp = 0 ;             // 方法中的参数，是局部变量
        temp = x + y ;               // 执行加法操作
        return temp ;                // 返回计算结果
    }
}
```

运算结果如下：

```
演示方法的调用！
addMethod1 的计算结果：30
addMethod2 的计算结果：23.6
```

递归调用是一种特殊的调用形式，属于方法的自身调用，在开发中应尽量避免使用。因为递归调用在操作时如果处理不好，就可能会出现内存的溢出，所以要谨慎使用，本书不再讲述。

2. 方法的重载

方法的重载就是方法名称相同，但参数类型或个数不同。通过传递参数的个数及类型的不同可以完成不同功能的方法调用。System.out.println() 方法就属于方法的重载，println() 方法可以打印数值、字符、布尔类型等数据。

下面的例子验证了方法的重载。

```java
public class MethodDemo {
    final static float PI = 3.14f;
    public static void main(String[] args) {
        int r = 3, a = 4, b = 5;
        System.out.println(" 圆形面积为：" + area(r));
        System.out.println(" 矩形面积为：" + area(a,b));
        System.out.println(" 三角形面积为：" + area(a,b,r));
    }
    public static float area(int r){// 定义 area 方法，完成圆面积的计算
        return PI*r*r;                              // 返回结果
    }
    // 定义 area 方法，完成矩形面积的计算
    public static int area(int a,int b){
        return a*b;                                 // 返回结果
    }
    // 定义 area 方法，完成三角形面积的计算
    public static double area(int a,int b,int r){
        double p = 0 ;
        p = (a+b+r)/2 ;
        return Math.sqrt(p*(p-a)*(p-b)*(p-r));  // 返回结果
    }
```

}

运算结果如下:

```
圆形面积为：28.26
矩形面积为：20
三角形面积为：6.0
```

 方法的重载只是在参数的类型或个数上有所不同,与方法的返回值无关。如果参数类型和个数一致,返回值不同,就不是方法的重载,并且在编译时无法通过,因为编译器无法判断是哪个方法。

3. 方法的引用传递

前面的操作传递和返回的都是基本数据类型。方法可以传递引用数据类型,数组属于引用数据类型,在把数组传递进方法之后,如果方法对数组本身做了任何修改,那么修改结果也将保存下来。下面的例子演示了传递数值和传递引用的不同。

```java
public class MethodDemo02 {
    public static void main(String[] args) {
        int x = 3, y = 4;
        change(3, 4);                          // 传递整型数值
        System.out.println("x=" + x + " y=" + y);
        int[] a = {3,4};
        change(a);                             // 传递数据引用
        System.out.println("a[0]=" + a[0] + " a[1]=" + a[1]);
    }
    public static void change(int x, int y) {
        x = x + y;
        y = x - y;
        x = x - y;
    }
    public static void change(int[] a) {
        a[0]=a[0]+a[1];
        a[1]=a[0]-a[1];
        a[0]=a[0]-a[1];
    }
}
```

运算结果如下:

```
x=3 y=4
a[0]=4 a[1]=3
```

从运行结果可以看出,基本数据类型传递的是数据的拷贝,而引用类型传递的是引用的复制。

2.6 要点总结

本章首先对 Java 的程序结构进行分析，讲解了 Java 的基本语法、原始数据类型及各类型之间转换的规则；然后介绍了 Java 的运算符、表达式及流程控制；最后讲解了数组与方法的定义和使用。

2.7 编程练习

1. 绘制柱状图：读入 5 个数，每个数介于 1~15，每个 * 代表一个数字，显示出柱状图。例如：

```
5, 12, 7, 10, 8
*****
************
*******
**********
********
```

2. 读入一个月份，假设该月份 1 日是周 3，请显示该月星期历，如果是 2 月份，就则认为是 28 天。

```
 0    1    2    3    4    5    6
                1    2    3    4
 5    6    7    8    9   10   11
12   13   14   15   16   17   17
19   20   21   22   23   24   25
26   37   28   29   30
```

3. 在排序好的数组中添加一个数字，并将添加后的数字插入到数组合适的位置。

第 3 章
Java 面向对象编程

前面学习的是 Java 的基本程序设计，属于结构化的程序开发。结构化方法的本质是功能分解，围绕实现处理功能的"过程"来构造系统。面向对象技术体现软件技术的多变性，具有稳定、可修改和可重用性的特点，可以很好地适应用户的变化。

3.1 理解面向对象

本节主要介绍 Java 面向对象的基本概念和特性。

3.1.1 基本概念

1. 对象

因为一个对象由一组属性和对这组属性进行操作的一组服务组成，所以它是属性和操作的封装体。对象是在面向对象系统中用于描述客观事物的一个实体，亦是构成系统的一个基本单位。

2. 类

从某种角度可以把类理解成对象的类型。类是具有相同属性和服务的一组对象的集合，它为属于该类的所有对象提供了统一的抽象描述。类与对象的关系就如同模具和铸件的关系，类的实例化结果就是对象，而对一类对象的抽象就是类。

3.1.2 基本特性

1. 封装性

封装性就是把对象的属性和服务组成对外相对独立而完整的单元。

对外相对独立是指对外隐蔽内部细节，只提供必要而有限的接口与外界交互。完整是指把对象的属性和服务结合在一起，形成一个不可分割的独立单位。

2. 继承性

继承是复用的重要手段。在继承层次中，高层的类相对于底层的类更抽象，更具有普遍性，如交通工具和汽车、火车、飞机的关系。交通工具处于继承层次的上层，相对于下层的汽车、火车或飞机等具体交通工具更为抽象和一般。由于汽车、火车和飞机除了体现交通工具的特性以外，又各自有不同的属性，提供的服务也各不相同，因此它们比交通工具更具体、更特殊。在 Java 中，通常把像交通工具这样抽象的、一般的类称作父类或超类，把像汽车、火车或飞机这样具体的、特殊的类称作子类。

3. 多态性

多态性是指在一般类中定义的属性或行为，被特殊类继承之后，可以具有不同的数据类型或表现出不同的行为。仍以交通工具和汽车、火车、飞机为例，交通工具都有驾驶的方法，虽然继承自交通工具的汽车、火车和飞机也同样具有驾驶的方法，但是它们具体驾驶的方法却不尽相同。

3.2 类与对象

Java 面向对象的核心组成是"类（class）"。本节介绍类的定义、类的对象的创建和使用，以及 this 和 static 关键字。

3.2.1 类定义

从类的概念中可以了解，类是由属性和方法组成的。属性中定义的是类需要的一个个具体信息，实际上一个属性就是一个变量，而方法就是一些操作的行为。Java 中类的定义形式如下：

```
[类修饰词] class 类名 [extends 超类名] [implements 接口列表]
{
    声明成员变量；              // 类的属性
    成员方法（函数）{};         // 定义方法的内容
}
```

定义 Student 类如下：

```java
public class Student {
    String name;                         // 声明姓名属性
    int age;                             // 声明年龄属性
    public void getStuInfo(){            // 取得学生信息的方法
        System.out.println("姓名: " + name + "年龄: " + age );
    }
}
```

类名应遵循标识符的命名规则，只是习惯类名的首字母大写。上面的类定义中还涉及类修饰词、接口、构造方法、成员方法等内容。

类修饰词也称作访问说明符。在上面的类定义中使用了类修饰词 public，除此之外，类修饰词还有 abstract、final 和默认类修饰词。类修饰词限定了访问和处理类的方式。

- public：被 public 修饰的类对所有类都是可见的，并且必须定义在以该类名为名字的 Java 文件中。
- final：被 final 修饰的类不能被继承，或者说不能作为超类也不可能有子类，这样的类编译器会对其进行优化。
- abstract：被 abstract 修饰的类是抽象类。因为抽象类至少有一个成员方法需要在其子类中给出完整定义，所以抽象类不能被实例化。
- 默认：如果没有指定类修饰词，就表示使用默认类修饰词。在这种情况下，其他类可以继承此类，同时在同一个包下的类可以访问引用此类。

类的成员变量与前面提到的变量用法没有差别。

类成员变量的修饰词分为两类：访问控制修饰词和非访问控制修饰词。

访问控制修饰词包括 private、protected、public 和默认。

- private：被 private 修饰的成员变量只对成员变量所在类可见。
- protected：被 protected 修饰的成员变量对成员变量所在类、该类同一个包下的类和该类的子类可见。
- public：被 public 修饰的成员变量对所有类可见。
- 默认：如果没有指定访问控制修饰词，就表示使用默认修饰词。在这种情况下，成员变量对成员变量所在类和该类同一个包下的类可见。

非访问控制修饰词包括 static、final、transient 和 volatile。

- static：被 static 修饰的成员变量仅属于类的变量，而不属于任何一个具体的对象。静态成员变量的值是保存在类的内存区域的公共存储单元，而不是保存在某一个对象的内存区间。任何一个类的对象访问它时，取到的都是相同的数据；任何一个类的对象修改它时，也都是对同一个内存单元进行操作。

- final：被 final 修饰的成员变量在程序的整个执行过程中都是不变的，可以用它来定义符号常量。
- transient：被 transient 修饰的成员变量是暂时性变量。Java 虚拟机在存储对象时不存储暂时性变量。在默认情况下，类中所有变量都是对象永久状态的一部分，当对象被存档时，这些变量同时被保存。
- volatile：被 volatile 修饰的成员变量不会被编译器优化，这样可以减少编译的时间。这个修饰词并不常用。

类的成员方法的命名必须是合法的标识符，一般是用于说明方法功能的动词或动名词短语。返回值类型可以是 void 和所有数据类型。若需要传入参数，则参数的定义包括参数类型和参数名；若需要一个以上的参数，则将不同的参数之间用逗号隔开形成参数列表。参数列表中的参数名不能相同。

在 Java 中，参数传递只有一种形式—传值。传值是指当参数被传递给一个方法时，方法中使用的是原始参数的副本。对于原始类型和引用类型都是如此。

类成员方法的修饰词也分为两类：访问控制修饰词和非访问控制修饰词。

访问控制修饰词包括 private、protected、public 和默认。

- private：被 private 修饰的成员方法只对成员方法所在类可见。
- protected：被 protected 修饰的成员方法对成员方法所在类、该类同一个包下的类和该类的子类可见。
- public：被 public 修饰的成员方法对所有类可见。
- 默认：如果没有指定访问控制修饰词，就表示使用默认修饰词。在这种情况下，成员方法对成员方法所在类和该类同一个包下的类可见。

非访问控制修饰词包括 static、final、abstract、native 和 synchronized。

- static：被 static 修饰的成员方法称作静态方法。静态方法是属于整个类的类方法，而不使用 static 修饰、限定的方法是属于某个具体类对象的方法。由于 static 方法是属于整个类的，因此不能操纵和处理属于某个对象的成员变量，只能处理属于整个类的成员变量。
- final：被 final 修饰的成员方法不会被子类继承。
- abstract：被 abstract 修饰的成员方法称作抽象方法。抽象方法是一种仅有方法头，没有方法体和操作实现的一种方法。
- native：被 native 修饰的成员方法称作本地方法。本地方法的方法体可以用像 C 语言这样的高级语言编写。
- synchronized：被 synchronized 修饰的成员方法用于多线程之间的同步。这个修饰词在后面章节中会有更详细的说明。

3.2.2 对象的创建及使用

1. 对象的创建

Java 中通过使用 new 关键字产生一个类的对象，这个过程也称作实例化。要想使用一个类，必须创建对象，其格式如下：

```
类名  对象名称 = null;                    // 声明对象
对象名称 = new 类名();                    // 实例化对象
```

也可以一步完成：

```
类名 对象名称（引用变量）= new 类名();
```

下面的例子是为上面定义的 **Student** 类创建对象。

```java
public class ClassDemo01 {
    public static void main(String[] args) {
Student student = new Student();              // 创建一个 student 对象
        student.name = "张三";                 // 设置 student 对象的属性内容
        student.age = 20;                      // 设置 student 对象的属性内容
        System.out.println(student.getStuInfo());
        Student student1 = new Student();// 创建一个 student 对象
        student1.name = "李四";                // 设置 student 对象的属性内容
        student1.age = 23;                     // 设置 student 对象的属性内容
        System.out.println(student1. getStuInfo ());
    }
}
class Student{
    String name;                               //学生姓名 -- 类的属性
    int age;                                   //学生年龄 -- 类的属性
    public String getStuInfo(){                //获取学生信息 -- 类的方法
        return  "学生姓名:" + name + "\t 学生年龄:" + age;
    }
}
```

从上面的示例中可以看到，访问对象中属性和方法的格式如下：

- 访问属性：对象名称.属性名。
- 访问方法：对象名称.方法名()。

2. 封装性

类的封装是指属性的封装和方法的封装。封装的格式如下：

- 属性封装：private 属性类型 属性名称。
- 方法封装：private 方法返回值 方法名称（参数列表）{}。

方法封装在实际开发中很少使用。下面的示例是为程序加上封装属性。

```java
public class ClassDemo02 {
    public static void main(String[] args) {
        Student student = new Student();        // 创建一个 student 对象
        student.name = "张三";                  // 错误，无法访问封装属性
        student.age = 20;                       // 错误，无法访问封装属性
        System.out.println(student.getStuInfo());
    }
}
class Student{
    private String name;                        // 学生姓名 -- 类的属性
    private int age;                            // 学生年龄 -- 类的属性
    public String getStuInfo(){                 // 获取学生信息 -- 类的方法
        return "学生姓名：" + name + "\t学生年龄：" + age;
    }
}
```

上面程序在编译时提示错误为"属性是私有的"。在 Java 开发中，对私有属性的访问有明确的定义："只要是被封装的属性，就必须通过 setter 和 getter 方法设置和取得。"

下面的示例是为前面类中私有属性加上 setter 和 getter 方法。

```java
public class ClassDemo03 {
    public static void main(String[] args) {
        Student student = new Student();        // 创建一个 student 对象
        student.name = "张三";                  // 设置 student 对象的属性内容
        student.age = 20;                       // 设置 student 对象的属性内容
        System.out.println(student.getStuInfo());
    }
}
class Student{
    String name;                                // 学生姓名 -- 类的属性
    int age;                                    // 学生年龄 -- 类的属性
    public int getAge() {                       // 取得年龄
        return age;
    }
    public void setAge(int age) {               // 设置年龄
        this.age = age;
    }
    public String getName() {                   // 取得姓名
        return name;
    }
    public void setName(String name) {          // 设置姓名
        this.name = name;
```

```
    }
    public String getStuInfo(){                    // 取得信息的方法
            return  "学生姓名:"+ name+"\t 学生年龄:"+ age;
    }
}
```

 开发中类的全部属性必须封装,通过 setter 和 getter 方法进行访问。

Eclipse 中【Source】--【Generate Setters and Getters】自动生成 setter 和 getter 方法。

3. 构造方法

实例化时,首先为对象分配内存,执行该类的构造方法,然后返回该对象的引用并将其赋给引用变量。类通过其定义构造方法产生对象。

把构造方法看作是一种特殊的类成员方法。构造方法的特殊性体现在以下两个方面:

(1) 构造方法的方法名必须与类名相同。

(2) 构造方法没有返回类型。

虽然构造方法有其特殊性,但也是类成员方法,所以构造方法可以重载。如果没有定义类的构造方法系统,就会自动提供一个默认的、无参数的构造方法。因为构造方法在类实例化时被调用,所以一般在方法体中初始化成员变量。下面是一个完整的类实例化示例。

```
class Employee {
    private String name;                    // 声明姓名属性
    private int salary;                     // 声明薪水属性
    Employee() {                            // 无参构造方法
        System.out.println(" 一个新的 Employee 对象产生 =========");
    }
    Employee(String name, int salary) {     // 有参构造方法
        this.setName(name);
        this.setSalary(salary);
    }
    Employee(int salary) {                  // 有参构造方法
        this.setSalary(salary);
    }
    public String getName() {               // 获得姓名
        return name;
    }
    public void setName(String name) {      // 设置姓名
        this.name = name;
```

```java
    }
    public int getSalary() {                    // 获得薪水
        return salary;
    }
    public void setSalary(int salary) {         // 设置薪水
        if (salary >= 0) {
            this.salary = salary;
        }
    }
}
public class ClassDemo03 {
    public static void main(String[] args) {
        System.out.println("声明一个对象Employee = null");
        Employee e = null;                      // 声明一个对象并不会调用构造方法
        // System.out.println("实例化对象：e = new Employee() ;");
        // e = new Employee();
        e = new Employee("sam", 3000);
        System.out.println("员工姓名：" + e.getName() + "\t员工工资："
+ e.getSalary());
        // new Employee("eva", 6000).getSalary();    // 匿名对象
    }
}
```

4. 匿名对象

匿名对象是指没有明确给出名称的对象。一般匿名对象只使用一次，并且只在椎内存中开辟空间，而不存在栈内存的引用，如上例中注释的匿名对象的使用。

3.2.3 this 和 static 关键字

1. this 关键字

Java 中 this 关键字语法比较灵活，主要有以下作用：

（1）表示类中的属性。

（2）调用本类的方法（成员方法和构造方法）。

（3）表示当前对象。

下面的示例演示了 this 的应用。

```java
public class ClassDemo04 {
    public static void main(String[] args) {
        Student s1=new Student("郭靖",23); // 声明两个对象，内容完全相等
        Student s2=new Student("郭靖",23); // 声明两个对象，内容完全相等
```

```java
            if (s1.compare(s2)) {
                System.out.println(" 是同一个学生！");
            } else {
                System.out.println(" 不是同一个学生！");
            }
        }
    }
}
class Student {
    private String name;                    // 声明姓名属性
    private int age;                        // 声明年龄属性
    public Student(){
        System.out.println(" 一个新的 Student 对象被实例化！");
    }
    public Student(String name, int age) {
        this();                 // 调用 Student 类的无参构造方法，必须放在第一行
        this.name = name;       // 表示本类中的属性
        this.age = age;
    }
    public int getAge() {                   // 取得年龄
        return age;
    }
    public String getName() {               // 取得姓名
        return name;
    }
    public boolean compare(Student stu){
        // 调用此方法时存在两个对象：当前对象，传入的对象 stu
        Student s1 = this;          // 表示当前调用方法的对象
        Student s2 = stu;           // 传递到方法中的对象
        if (s1 == s2) {             // 首先比较两个地址是否相等
            return true;
        }
        // 分别判断每一个属性是否相等
        if (s1.name.equals(s2.name)&&s1.age == s2.age) {
            return true;
        } else {
            return false;
        }
    }
    public void getStuInfo(){               // 取得学生信息
        // this 调用本类中的方法，如：getter 方法
        System.out.println(" 姓名: " + this.getName() + " 年龄: "
+ this.getAge());
    }
}
```

程序运行结果如下：

```
一个新的 Student 对象被实例化！
一个新的 Student 对象被实例化！
是同一个学生！
```

2. static 关键字

static 关键字声明的属性和方法称为类属性和类方法，被所有对象共享，直接使用类名称进行调用。例如：

```java
public class ClassDemo05 {
    public static void main(String[] args) {
        Student s1 = new Student(" 小李 ",23);        // 声明 Student 对象
        Student s2 = new Student(" 小王 ",30);        // 声明 Student 对象
        s1.getStuInfo();                              // 输出学生信息
        s2.getStuInfo();                              // 输出学生信息
        Student.grade = "09 级网络工程 ";             // 类名称调用修改共享变量的值
//      s1.grade = "09 级网络工程 ";                  // 对象也可以对共享变量赋值
        s1.getStuInfo();                              // 输出学生信息
        s2.getStuInfo();                              // 输出学生信息
    }
}
class Student {
    static String grade = "09 级软件工程 ";
    private String name;                              // 声明姓名属性
    private int age;                                  // 声明年龄属性
    public Student(String name, int age) {
        this.name = name;                             // 表示本类中的属性
        this.age = age;
    }
    public int getAge() {                             // 取得年龄
        return age;
    }
    public String getName() {                         // 取得姓名
        return name;
    }
    public void getStuInfo(){                         // 取得学生信息
        // this 调用本类中的方法，如 getter 方法
        System.out.println("姓名: " + this.getName() + "\t 年龄: "
            + this.getAge() + "\t 班级 :" + this.grade);
    }
}
```

程序运行结果如下：

姓名：小李	年龄：23	班级：09 级软件工程
姓名：小王	年龄：30	班级：09 级软件工程
姓名：小李	年龄：23	班级：09 级网络工程
姓名：小王	年龄：30	班级：09 级网络工程

读者还记得第 2 章分析了 Java 的 4 块内存区域吗？下面通过上例来分析一下程序执行过程的内存分配，如图 3-1 所示。

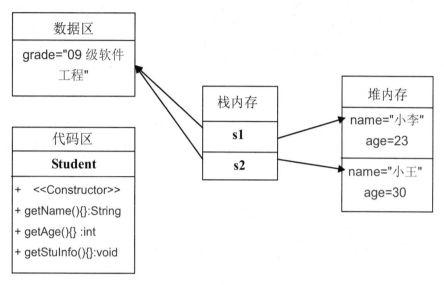

图 3-1 static 属性保存的内存分配

static 关键字声明方法称为类方法，由类直接调用，本身前面已经多次使用了 static 声明的方法。由图 3-1 可知，所有的方法都放在代码区，也是多个对象共享的内存区，但是非 static 声明的方法属于所有对象共享的区域，而 static 属于类，即不用实例化对象也可以通过类调用执行。static 声明的方法是不能调用非 static 声明的属性和方法的，反之则可以。例如：

```
public class StaticDemo {
    private static String grade = "09 软件工程 ";    // 定义静态属性
    private String name = "sam";                    // 定义私有成员变量
    public static void refFun(){
        System.out.println(name);                   // 错误，不能调用非 static 属性
        fun();                                      // 错误，不能调用非 static 方法
    }.
    public void fun(){
        System.out.println(" 非 static 方法！ ");
        refFun();                                   // 非 static 方法可以调用 static 方法
    }
    public static void main(String[] args) {
        refFun();                                   // static 方法可以调用 static 方法
        new StaticDemo().fun();                     // 通过实例化对象调用非 static 方法
```

```
        }
    }
```

通过编译出现错误的信息可知，static 是不能调用任何非 static 内容的，因为不知道非 static 的内容是否被初始化了，读者在开发中要谨慎对待。

3.2.4 内部类

在类的内部可以定义属性和方法，也可以定义另一个类，叫内部类。包含内部类的类叫外部类。内部类可声明 public 或 private，访问权限与成员变量、成员方法相同。内部类的方法可以访问外部类的成员，且不必实例化外部类，反之则不行。通过下面一个简单的例子了解内部类的使用。

```
class Outer{
    int temp = 10;                          // 外部类的属性
    String author = "sam";                  // 外部类的属性
    class Inner{                            // 内部类的定义
        int temp = 20;                      // 内部类的属性
        public void showOuter(){            // 内部类的方法
            // 外部类的调用
            System.out.println(" 外部类的author:"+author);
            System.out.println(" 内部类的temp:"+temp);
            System.out.println(" 外部类的temp:"+Outer.this.temp);
        }
    }
    public void showInner(){
        Inner in = new Inner();
        in.showOuter();
    }
}
public class InnerClassDemo01 {
    public static void main(String[] args) {
        Outer out = new Outer();
        out.showInner();
    }
}
```

程序运行结果如下：

```
外部类的author:sam
内部类的temp:20
外部类的temp:10
```

在内部类之外，还有一种匿名内部类，主要用于 GUI（图形用户界面）编程。在讲完接口和抽象类后再介绍其使用方法。

3.3 继承

在面向对象程序设计中,继承是不可或缺的一部分。通过继承可以实现代码的重用,提供程序的可维护性。

3.3.1 继承的语法和规则

图 3-2 描述了继承的关系。

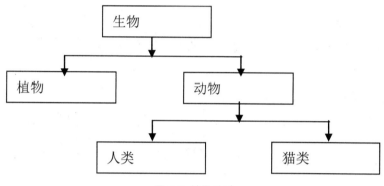

图 3-2 继承关系

图 3-2 中顶端是范围较大的类,向下详细分成几个小类,这样的分类关系称为继承关系,上面的大类为父类,下面的小类为子类。

父类也称基类、超类;子类也称衍生类。因为子类继承了父类的所有特征,同时子类在父类的基础上还增加了自己的特征,所以子类和父类相比具有更丰富的功能。

在继承关系中还能够发现一个规律:子类是父类的一种,也可以说"子类就是父类",如人类是动物,动物就是生物等,记住这个定律对理解继承的概念非常有帮助。

继承的语法格式如下:

[修饰符] class 子类名 extends 父类名

观察下面的例子理解继承的语法和规则。

```
public class ExtendsDemo01 {
    public static void main(String[] args) {
        Person p = new Person();           // 实例化父类对象
        p.name = "sam";                     // 父类对象的属性赋值
        p.age = 22;                         // 父类对象的属性赋值
        p.height = 1.76;                    // 父类对象的属性赋值
        Student s = new Student();          // 实例化子类对象
```

```java
            s.score = 83.0 ;                            // 子类对象的属性赋值
            System.out.println(" 子类的信息: " + s.name + "\t"
                    + s.age + "\t" + s.height + "\t" + s.score);
            s.sayHello();                               // 调用子类方法
        }
}
class Person {
    String name ;                   // 声明类 Person 的姓名属性
    int age ;                       // 声明类 person 的年龄属性
    double height ;                 // 声明类 person 的身高属性
    public Person(){
        System.out.println(" 父类的构造方法 ");
    }
    public void sayHello(){
        System.out.println(" 父类的方法 sayHello() 方法 ");
    }
}
class Student extends Person{
    double score ;                  // 声明子类 Student 的学分属性
    public Student() {
        System.out.println(" 子类的构造方法 ");
    }

    public void sayHello(){
        System.out.println(" 子类的 sayHello() 方法 ");
    }
}
```

程序运行结果如下:

```
父类的构造方法
父类的构造方法
子类的构造方法
子类的信息: null    0        0.0       83.0
子类的 sayHello() 方法
```

从程序运行结果分析,继承有以下特性:

(1)子类继承父类所有的属性和方法,同时也可以在父类继承上增加新的属性和方法。

(2)子类不继承父类的构造器。

(3)子类可以继承父类中所有的可被子类访问的成员变量和方法,但必须遵循以下规则:

- 子类不能继承父类声明为 private 的成员变量和成员方法;
- 如果子类声明了一个与父类成员变量同名的成员变量,子类就不能继承父类的成员变量，此时称子类的成员变量隐藏了父类的成员变量；
- 如果子类声明了一个与父类的成员方法同名的成员方法，子类就不能继承父类的成员方法，此时称子类的成员方法隐藏了父类的成员方法。

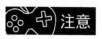 请读者务必在上面示例的基础上验证继承的语法和规则。

3.3.2 重载和覆盖

重载是指定义多个方法名相同但参数不同的方法。本书在第 2 章已经详细讲解了重载的规则和使用方法，这里不再赘述。覆盖也称覆写，是继承关系中方法的覆盖。方法覆盖需要满足以下规则。

（1）发生在父类和子类的同名方法之间。
（2）两个方法的返回值类型必须相同。
（3）两个方法的参数类型、参数个数、参数顺序必须相同。
（4）子类方法的权限必须不小于父类方法的权限 private<defult<public。
（5）子类方法只能抛出父类方法声明抛出的异常或异常子类。
（6）子类方法不能覆盖父类中声明为 final 或 static 的方法。
（7）子类方法必须覆盖父类中声明为 abstract 的方法（接口或抽象类）。

下面的示例演示了方法的覆盖。

```java
public class ExtendsDemo02 {
    public static void main(String[] args) {
        Dog dog = new Dog();
        dog.cry();           // 覆盖父类的方法
        Cat cat = new Cat();
        cat.cry();           // 覆盖父类的方法
        Cattle cattle = new Cattle();
        cattle.cry();        // 没有覆盖父类的方法
    }
}
class Animal {
    public Animal(){
        System.out.println("Animal 类的构造方法！ ");
    }
    public void cry(){
        System.out.println(" 动物发出叫声！ ");
    }
}
```

```
class Dog extends Animal{
   public Dog(){
         System.out.println("Dog 类的构造方法！");
   }
   public void cry(){
         System.out.println(" 狗发出 " 汪汪 ..." 叫声！ ");
   }
}
```

这段程序代码的运行结果如下：

```
Animal 类的构造方法！
Dog 类的构造方法！
狗发出 " 汪汪 ..." 叫声！
Animal 类的构造方法！
Cat 类的构造方法！
猫发出 " 喵喵 ..." 叫声！
Animal 类的构造方法！
动物发出叫声！
```

从以上结果看出，通过覆盖可以使一个方法在不同的子类中表现出不同的行为。

3.3.3 super 关键字

super 关键字代表当前超类的对象。super 表示从子类调用父类中的指定操作，如调用父类的属性、方法和无参构造方法，有参构造方法。如果调用有参构造方法，就必须在子类中明确声明。与 this 关键字一样，super 关键字必须在子类构造方法的第一行。

下面的示例演示了 super 关键字的使用方法。

```
public class ExtDemo03 {
   public static void main(String[] args) {
         Santana s = new Santana("red");
   }
}
class Car{
   String color;
   Car(String color){
         this.color = color;
   }
}
class Santana extends Car{
   private String color;
   public Santana(String color) {
         super(color);
```

```
    }
    public void print(){
         System.out.println(color);
         System.out.println(super.color);
    }
}
```

3.4 final 关键字

final 关键字的中文含义是"最终的",它可以修饰很多成员,因修饰的成员不同含义也不同。

3.4.1 final 变量

final 关键字修饰的变量可以分为属性、局部变量和形参。无论修饰哪种变量,其含义都是相同的,即变量一旦赋值就不能改变。例如:

```
public class FinalVar {
   public final static double PI = 3.14;      // 常量
   final int x = 100;
   public static void main(String[] args) {
        final int y = 0;
   }
   public static void add(final int z){
        z++ ;                                   // 错误
   }
}
```

3.4.2 final 方法

final 关键字也可以修饰方法,这样的方法不能被子类覆盖。例如:

```
class FinalMethod {
   public final void add(int x) {
        x++;
   }
}
public class Sub extends FinalMethod {
   public void add(int x) {
        x += 2;
   }
}
```

编译时提示错误。

3.4.3 final 类

final 关键字修饰的类不能被继承，也不能产生子类。例如：

```
class FinalClass {
   public void add(int x) {
        x++;
   }
}
public class Sub1 extends FinalClass {
   public void add(int x) {
        x += 2;
   }
}
```

编译时提示错误。

3.5 抽象类

被 abstract 修饰词修饰的类称为抽象类。抽象类包含抽象方法的类。

抽象方法：只声明未实现的方法，抽象类必须被继承。如果子类不是抽象类，就必须覆写抽象类中的全部抽象方法。

```
abstract class A {
   public final static String FLAG = "china";
   public String name = "sam";
   public String getName() {
        return name;
   }
   public void setName(String name) {
        this.name = name;
   }
   public abstract void print();  // 比普通类多了一个抽象方法
}

class B extends A{   // 继承抽象类，因为 B 是普通类，所以必须覆写全部抽象方法
   public void print() {
        System.out.println(" 国籍： " + super.FLAG);
        System.out.println(" 姓名： " + super.name);
   }
}
public class AbstractDemo01 {
```

```
       public static void main(String[] args) {
//         A a = new A();            // 因为有未实现的方法，所以不能被直接实例化
           B b = new B();
           b.print();
       }
}
```

抽象类是不完整的类，不能通过构造方法被实例化。但这不代表抽象类不需要构造方法，其构造方法可以通过前面介绍的 super 关键字在子类中调用。另外，从语法的角度来说，抽象类可以没有抽象方法，但如果类定义中声明了抽象方法，那么这个类必须声明为抽象类。

 不可实例化的类不需要构造方法的说法是错误的。

3.6 接口

接口是 Java 中的重要组成部分，是由常量和公共的抽象方法组成的。也可以理解为是更纯粹的抽象类，即接口是抽象方法和常量值的定义集合，只包含常量和方法的定义，没有变量和方法的实现。

3.6.1 接口定义

Java 中接口的定义形式如下：

```
[修饰词] interface 接口名
{
    常量声明
    方法声明
}
```

接口是另一种引用类型。接口的修饰词只有 public 和默认两个，含义与类修饰符相同。接口名的命名规则也与类名相同。因为接口中变量的修饰词只能是 public、final、static，所以不用显式地使用修饰词。又因为接口中变量的修饰词是 public、final、static，所以在接口中声明的都是常量。接口中的方法都没有方法体，除了定义的常量以外也没有变量。接口中方法的修饰词只能是 public，默认也是 public。

另外，接口也有继承机制并支持多继承，可以使用 extends 关键字继承其他接口。

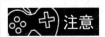 不同于类，接口支持多继承机制。

3.6.2 实现接口

接口只是声明了提供的功能和服务,而功能和服务具体的实现还是要在接口的类中定义。一个类可以通过 implements 关键字实现接口,接口在使用时也必须有子类,子类要实现(覆写)接口的所有抽象方法。一个子类可以同时实现多个接口,实现接口的类称为实现类。类之间实现继承,接口之间也可以实现继承,类和接口之间可以实现多接口的继承。

抽象类和接口实际上是一套规范,规定子类(实现类)必须定义的方法,除非子类(实现类)严格执行了这套规范,否则将不能实例化和使用。

下面的示例演示了计算机主板在工作时接口的实现。

```java
interface VideoCard{                  // 显卡接口
    void display();                   // 显卡工作的抽象方法
    String getName();                 // 获取显卡厂商名字的抽象方法
}
class Dmeng implements VideoCard{     // 具体厂商的显卡
    private String name;
    Dmeng(){
        name = "Dmeng's videoCard";
    }
    public void setName(String name) {
        this.name = name;
    }
    public String getName() {
        return this.name;
    }

    public void display() {
        System.out.println("Dmeng's videoCard working!!!");
    }
}
class Mainboard{
    private String CPU;
    VideoCard vc;
    public String getCPU() {
        return CPU;
    }
    public void setCPU(String cpu) {
        CPU = cpu;
    }
    public VideoCard getVc() {
        return vc;
    }
```

```java
        public void setVc(VideoCard vc) {
            this.vc = vc;
        }
        public void run(){
            System.out.println(CPU);
            System.out.println(vc.getName());
            vc.display();
            System.out.println("Mainboard's running!!!");
        }
}
public class Computer {
    public static void main(String[] args) {
        Dmeng dm = new Dmeng();
        Mainboard mb = new Mainboard();
        mb.setCPU("Intel's CPU");
        mb.setVc(dm);
        mb.run();
    }
}
```

3.6.3 匿名内部类

匿名内部类就是没有名字的内部类，经常被应用于 Swing 程序设计中的事件监听处理。例如创建一个匿名的内部类 ButtonAction：

```java
public class ClassDemo {
    public static void main(String[] args) {
        new ButtonAction(){
            public void click(){
                System.out.println("这是匿名类,但谁也无法使用它! ");
            }
        }
    }
}
```

匿名类通常用于创建接口的唯一实现类，或者创建某个类的唯一子类。

3.7 包及访问控制权限

包是 Java 中的文件组织形式，对应系统的文件夹。一个文件夹下可以存在文件也可以包含另一个文件夹，包也是如此。正因为有了包的存在，Java 工程中允许存在同名不同包的 Java 文件，所以指定一个类时除了类名还要有类所在的包路径。

3.7.1 包的操作

Java 中包的声明形式如下:

```
package 包名
```

例如:

```
package example.code.oo;
```

包名一般都为小写。包像文件夹一样可以嵌套，上面的例子表示 oo 包在 code 包下，而 code 包又存在于 example 包中。

一个类如果要引用其他包下的类，就需要使用 import 关键字。而指定一个类，就需要指定这个类所在包的完全路径。例如:

```
import java.util.ArrayList;
```

 虽然可以使用 * 代表包下所有的类，但不建议使用，建议完整指定要引入的类的路径。

3.7.2 访问权限修饰符

前面已经介绍过访问权限修饰符，下面详细讲解 4 种权限的不同。按照权限大小排序:

```
public > protected > default > private
```

1. 权限修饰符 public

最大的权限，完全开放，没有任何限制。任何类都可以调用 public 权限的方法，都可以访问 public 权限的属性。构造方法和类的权限通常为 public。

2. 权限修饰符 private

最小的权限，限制类外的访问，它修饰的成员对类来说是私有的。一般情况下，把属性设置为 private，让其他类不能直接访问属性，达到保护属性的目的。

3. 默认权限修饰符

该权限修饰的成员在类内可以访问，同一个包内的其他类也可以访问，其他包中的类不能访问。

4. 权限修饰符 protected

该权限修饰的成员能够被子类和同一个包中的类访问。

以上 4 个权限修饰符可以修饰类的成员，不能修饰局部变量。

3.8 对象的多态性

多态性在面向对象中是一个非常重要的概念，主要有以下两种形式：

（1）方法的重载和覆盖。

（2）对象的动态性。

方法的重载和覆盖可参看 3.3.2 节，下面重点介绍对象的多态性。对象的多态性主要分为以下两种类型：

（1）向上转型：子类对象－＞父类对象。

（2）向下转型：父类对象－＞子类对象。

对于向上转型，程序会自动完成，而对于向下转型，必须明确指明要转型的子类类型。格式如下：

对象向上转型：父类 父类对象 = 子类对象

对象向下转型：子类 子类对象 =（子类）父类对象

示例代码如下：

```
class Person {
  private String name;
  private int age;
  Person(String name,int age){
      this.name = name;
      this.age = age;
  }
  public String toString() {
      return " 姓名："+name+"，年龄 "+age;
  }
}
class Teacher extends Person {
  private float salary;
  Teacher(String name,int age,float salary) {
      super(name, age);
      this.salary = salary;
  }
  public String toString() {
      return super.toString()+"，薪水 "+salary;
  }
}
```

```java
class Student extends Person {
    private float score;
    Student(String name, int age ,float score) {
        super(name, age);
        this.score = score;
    }
    public String toString() {
        return super.toString()+",学生成绩："+score;
    }
}
public class PolDemo02 {
    public static void main(String[] args) {
        Person p = new Teacher("eva",33,2000.0f);         // 向上转型
        Teacher t = (Teacher)p;                             // 向下转型
        System.out.println(p.toString());
        System.out.println(t.toString());
        /*Person p = new Person("john",30);
        Teacher t = (Teacher)p;
        System.out.println(p.toString());
        System.out.println(t.toString());*/
    }
}
```

程序运行结果如下：

姓名：eva，年龄 33，薪水 2000.0
姓名：eva，年龄 33，薪水 2000.0

程序中有一程序块的注释是不能实现的，读者可以去掉注释运行一下，会提示"错误的转型"。在进行对象的向下转型前，必须先发生对象的向上转型，否则会出现对象转换异常。也就是说，对象不允许不经过向上转型而直接向下转型。

经过向上和向下转型后，可能会出现某个引用到底指向哪种类型对象，可以使用 instanceof 关键字判断一个对象到底是哪个类的实例。格式如下：

对象引用 instanceof 类名   ->   返回 boolean 类型
对象引用 instanceof 接口名   ->   返回 boolean 类型

请读者自行测试。

3.9 Object 类

在 Java 中，所有的类都有一个公共的父类 Object。一个类只要没有明显地继承一个类，

就肯定是 Object 类的子类。Object 是 Java 中唯一一个没有父类的类，是 Java 最顶层的父类。Object 类中的主要方法如表 3-2 所示。

表3-2 Object类中的主要方法

序号	方法名称	类型	描述
1	public Object()	构造	构造方法
2	public boolean equals(Object obj)	普通	对象比较
3	public String toString()	普通	对象输出
4	public int hashCode()	普通	取得hash码

1. toString() 方法

观察下面程序的运行结果。

```
class Student{
  private String name;
  private int age;
  public Student(String name,int age){
        this.name = name;
        this.age = age;
  }
}
public class ObjectDemo {
  public static void main(String[] args) {
        Student stu = new Student("sam",20);
        System.out.println(stu);
        System.out.println(stu.toString());
  }
}
```

运行结果如下：

```
Student@1c78e57
Student@1c78e57
```

从运行结果可以看出，加不加 toString() 方法输出结果都一样，即对象输出时一定会调用 Object 类中的 toString() 方法。通常情况下，toString() 方法应该返回能够简明扼要地描述对象的文本，而上面的字符串不包含有意义的描述信息。所以，一般子类都会覆盖该方法，让其返回有意义的文本。例如在 Student 类中覆盖 toString() 方法的代码如下：

```
public String toString() {
        return "姓名：" + name + " 年龄：" + age;
}
```

运行结果变为：

姓名：sam 年龄：20

2. equals() 方法

因为 Object 类提供的 equals() 方法默认是比较地址的，并不能对内容进行比较，所以自定义的类如果比较内容，就需要覆盖 Object 类的 equals() 方法。

方法 equals() 与运算符"=="都是判断是否相等，返回 boolean 值，使用时很容易混淆，下面加以区别。

（1）使用范围不同

运算符"=="可以比较基本数据类型，也可以比较引用数据类型，而 equals() 方法只能比较引用数据类型。

（2）功能不同

运算符"=="的功能

- 比较基本数据类型：相当于比较数值是否相等；
- 比较引用数据类型：比较两个引用的值，即地址是否相等。

equals() 方法的功能

比较两个引用的值，即地址是否相等。很多子类中的 equals() 方法覆盖了 Object 类的 equals() 方法，改变了方法的功能，如 String 类等。

3. hashCode() 方法

该方法返回对象的哈希码。哈希码是一个代表对象的整数，可以把哈希码比作对象的身份证号码。在程序运行期间，每次调用同一对象的 hashCode() 方法，返回的哈希码必定相同。但是多次执行同一个程序，程序的一次执行和下一次执行期间，同一对象的哈希码不一定相同。实际上，默认的哈希码是将对象的内存地址通过某种转换得到的，因此不同对象会有不同的哈希码。

3.10 包装类

Java 的设计思想是一切皆为对象，但 Java 的 8 种基本类型并不是对象，因此 Sun 为 8 种基本数据类型分别增加了属性和方法，生成了相对应的 8 个类，称为包装类（Wrapper）。具体见表 3-3。

表3-3 包装类

序号	基本数据类型	包装类
1	Byte	java.lang.Byte
2	Short	java.lang.Short

(续表)

序号	基本数据类型	包装类
3	Int	java.lang.Integer(注意类名)
4	Long	java.lang.Long
5	Float	java.lang.Float
6	Double	java.lang.Double
7	Char	java.lang.Character(注意类名)
8	Boolean	java.lang.Boolean

由表 3-3 可知，每个包装类都有一些类似功能的方法，下面介绍常用的方法。

3.10.1 基本数据类型转换为包装类

包装类就是把基本数据类型进行包装的类。基本数据类型会被包装起来成为类的一个属性，同时还会增加一些新的属性和方法。

```
int x = 6 ;
Integer y = new Integer(x) ;
Integer z = new Integer(12) ;
```

整型类型变量 x 和 12 分别被包装为 Integer 包装类对象 y 和 z，使 y 和 z 具有了属性和方法，这些属性和方法能够把包装的整数转换为各种形式。帮助文档中 8 个包装类都有对应的构造方法把基本数据类型转换为包装类。

3.10.2 字符串转换为包装类

字符串转换为包装类，也要使用包装类的构造方法。

```
String s = "3.14";
Double d = new Double(s);
Boolean b = new Boolean("true");
```

上面 3 个字符串通过构造方法转换为相应的包装类。除了 character 包装类，其他 7 个都有把字符串转换为包装类的构造方法。如果字符串不是合法的数字组成，就会转换不成功，运行时会抛出 NumberFormatException 类异常。

3.10.3 包装类转换为基本数据类型

基本数据类型转换为包装类之后，增加了新的属性和方法。但是因为转换为包装类之后，包装类就不能像基本数据类型那样参与运算，被包装的值不能改变，所以如果进行运算的话，就需要把包装类转换为基本数据类型。格式如下：

```
public type typeValue()
```

其中 type 代表 8 个基本数据类型,如 intValue()、floatValue() 等。例如:
```
Integer x = new Integer(3);
Integer y = new Integer(6);
int a = x. intValue();
int b = y. intValue();
int c = a + b;
```

3.10.4 字符串转换为基本数据类型

7 个包装类都有静态方法可以实现字符串转换为基本数据类型。格式如下:

```
public static type parsetype (String s)
```

其中 type 代表除字符串外 7 个基本数据类型,如 parseInt()、parseFloat() 等。如果字符串不是合法的数字组成,运行时就会抛出 NumberFormatException 类异常。

各种数据类型的转换关系如图 3-3 所示。

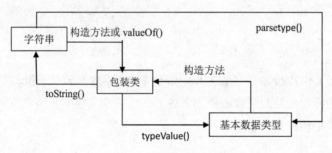

图 3-3 数据类型的转换

3.10.5 自动装箱和自动拆箱

基本数据类型转换为包装类称为装箱,包装类转换为基本数据类型称为拆箱。装箱时使用的是构造方法,拆箱时可以使用 parsetype() 方法。在 JDK5.0 版本以后,简化了装箱和拆箱的过程,使用了自动机制,使装箱和拆箱的编码更为简单。例如:

```
public class Demo {
    public static void main(String[] args) {
        Integer x = 10 ;
        Integer y = 20 ;
        Integer z = x + y ;
        System.out.println(z);
    }
}
```

实际上这些装箱和拆箱的工作在编译器完成。

3.10.6 覆盖父类的方法

包装类是 Object 的子类，在包装类里覆盖了父类的方法，常用的有 equals() 和 toString() 方法。

覆盖后的 equals() 方法不再比较引用的值，而是比较被包装的基本数据类型的值是否相等。覆盖后的 toString() 方法返回被包装的基本数据类型的值。

3.11 String 类

String 是用于声明字符串的类，在 Java 中是一个特殊的类。请读者详细参考 JDK API，了解 String 类的作用和相关操作。

3.11.1 String 对象的实例化和内容比较

String 类有两种实例化对象的方法。因为 String 是类，所以有类的实例化对象的方法。通过下面的示例理解 String 类的实例化对象的方法。

```java
public class StringDemo01 {
    public static void main(String[] args) {
        // 直接实例化 String 对象
        String s1 = "ChenZhanWei";
        String s2 = "ChenZhanWei";
        // 调用 String 类中的构造方法实例化对象
        String s3 = new String("ChenZhanWei");
        String s4 = new String("ChenZhanWei");
        // "==" 比较
        System.out.println("s1==s2->" + (s1==s2));
        System.out.println("s3==s4->" + (s3==s4));
        System.out.println("s1==s3->" + (s1==s3));
        // String 的内容比较
        System.out.println("s1 equals s2->" + (s1.equals(s2)));
        System.out.println("s3 equals s4->" + (s3.equals(s4)));
        System.out.println("s1 equals s3->" + (s1.equals(s3)));
        // 修改字符串的内容
        String s1 = "JKX->" + s1;
        System.out.println(s1);
    }
}
```

程序运行结果如下：

```
s1==s2->true
```

```
s3==s4->false
s1==s3->false
s1 equals s2->true
s3 equals s4->true
s1 equals s3->true
JKX->ChenZhanWei
```

从程序运行结果中可以发现：

（1）String 类的 equals() 方法重写了 Object 类的 equals() 方法。

（2）String 类的两种实例化方法在实例化对象时存在差别。原因如下：

Java 中提供了一个字符串池来保存全部内容，这是 Java 的共享设计，即以直接赋值的方式声明的多个对象在一个对象池中。新实例化的对象如果在池中已经定义了，就不再重新定义，而是直接使用，即对象 s1、s2 指向对内存中字符串池中的同一个对象。如果使用 new 关键字，无论如何都会开辟一个新的空间，对象 s3、s4 实际上是开辟了两个内存空间的引用，其地址值是不同的。建议使用直接实例化对象的方法。

（3）String 类可以修改字符串的内容。

实际上字符串的内容是不可以更改的，下面通过内存分配图理解字符串内容的不可更改性，如图 3-4 所示。

图 3-4 字符串修改的内存分配

从上图可知，一个 String 对象内容的改变实际上是内存地址的改变，而本身字符串的内容并没有改变。字符串的修改一般由 StringBuffer 类完成。

3.11.2 String 类中的常用方法

在 String 类中提供了大量的操作方法，读者可参考 JDK API 文档，本书只介绍常用的几个方法。示例如下：

```java
public class StringDemo02 {
    public static void main(String[] args) {
        String s = " zknu.edu.cn " ;
        char[] c = s.toCharArray();        // 将字符串变为字符数组
        for (int i = 0; i < c.length; i++) {
```

```
                    System.out.print(c[i] + "\t");          // 循环输出
            }
            System.out.println();                            // 换行
            String s1 = new String(c);              // 将全部字符串数组变为String
            String s2 = new String(c,6,3);          // 将部分字符串数组变为String
            System.out.println(s1);                          // 输出字符串
            System.out.println(s2);                          // 输出字符串
            System.out.println(s.charAt(6));                 // 取出字符串中第 6 个字符
            System.out.println(s.length());                  // 取得字符串的长度
            // 查找指定字符串是否存在，返回位置
System.out.println(s.indexOf("n"));
// 从字符串的指定位置查找，返回位置，没有找到返回 -1
System.out.println(s.indexOf("k",6));
            System.out.println(s.trim());                    // 去掉左右空格后输出
            System.out.println(s.substring(1, 5));  // 字符串截取
    }
}
```

程序运行结果如下：

```
    z   k   n   u   .   e   d   u   .   c   n
zknu.edu.cn
edu
e
13
3
-1
zknu.edu.cn
zknu
```

String 类中字符串的比较、大小写转换、内容的替换及拆分字符串、判断是否以指定的字符串开头或结尾在开发中也经常用到，请读者自行练习，本书不再赘述。

3.12 要点总结

本章首先介绍了面向对象的基本概念和特性；然后分别对 Java 中类和对象、继承、抽象类、接口和包这三个重要概念进行详细介绍，同时讲解设计类的三个特性：封装、继承和多态的概念和使用；最后介绍了 Object 类、包装类和 String 类。

3.13 编程练习

1. 编写一个程序，得出三个不同盒子的体积。将每个盒子的高度、宽度和长度参数的

值传递给构造方法，计算并显示体积。

2. 编写一个程序，显示水果的定购详情。定义一个带有参数的构造方法，这些参数用于存放产品名、数量和价格。此程序接受并输出构造方法的参数，即三种不同的水果。

3. 编写一个程序，用于创建一个名为 Employee 的父类和两个名为 Manager 和 Director 的子类。Employee 类包含三个属性和一个方法，属性为 name、basic 和 address，方法为 show()，用于显示这些属性的值。Manager 类有一个称为 department 的附加属性，Director 类有一个称为 transportAllowance 的附加属性，创建 Manager 和 Director 类的对象并显示其详细信息。

4. 编写一个程序，用于重写父类 Addition 中名为 add() 的抽象方法。add() 方法在 NumberAddition 类中将两个数字相加，在 TextConcatenation 类中连接两个 Strings（字符串）。声明属性并在父类 Addition 的构造方法中初始化属性。

第 4 章
Java 异常

异常是在运行时程序代码序列中产生的一种异常情况、异常事件，如文件读写时找不到指定的路径、数据库操作时连接不到指定的数据库服务器等，此时程序无法继续运行，导致整个程序运行中断。本章主要介绍 Java 程序设计中异常的概念及处理机制。

4.1 Java 中的异常类及分类

Java 提供了一套完整的异常处理机制。

在 Java 中，一切的异常都秉承着面向对象的设计思想，所有的异常都以类和对象的形式存在。除了 Java 中已经提供的各种异常类外，用户也可以根据需要自定义异常类。

在程序实际的应用中，任何程序都可能存在问题，所以在程序的开发中对于错误的处理是极其重要的，而 Java 提供的异常处理机制就可以帮助用户很好地解决这些问题。

 引入异常机制并不代表之后的程序设计将完全摒弃传统的错误。异常机制也有自身的不足，正确的做法是将两者互补性地结合起来使用。

Java 通过面向对象的方法进行异常处理，把各种不同的异常进行分类，并提供良好的接口。在 Java 中，每个异常都是一个对象，它是 Throwable 类或其子类的实例。当一个方法出现异常后便抛出一个异常对象，此时系统（JVM）会自动实例化一个异常类对象，该对象中保存了具体的异常描述信息，调用这个对象的方法可以捕获到这个异常并进行处理。因

此，在这里有必要先了解一下 Throwable 类。

Throwable 类的 4 个构造方法如下：

```
Throwable()
Throwable(String message)
Throwable(String message, Throwable cause)
Throwable(Throwable cause)
```

Throwable 类的方法如下：

```
Throwable  fillInStackTrace()
Throwable  getCause()
String  getLocalizedMessage()
String  getMessage()
StackTraceElement[]  getStackTrace()
Throwable  initCause(Throwable cause)
void  printStackTrace()
void  printStackTrace(PrintStream s)
void  printStackTrace(PrintWriter s)
void  setStackTrace(StackTraceElement[] stackTrace)
String  toString()
```

Throwable 类及其子类关系如图 4-1 所示。

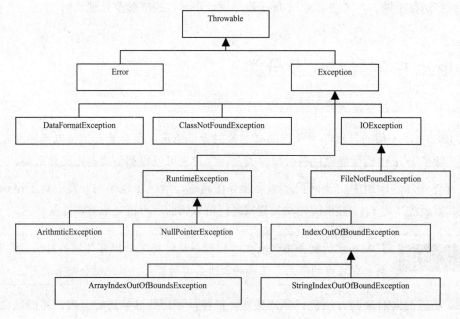

图 4-1 Throwable 类及其子类关系

Throwable 类有两个重要的子类：Error 和 Exception。

Error 类特指应用程序在运行期间发生的严重错误，如虚拟机内存用尽、堆栈溢出、动态链接失败等,对于这类错误导致的应用程序中断,程序无法预防和恢复。一般情况下,这种错误都是灾难性的,没有必要使用异常处理机制处理。

Exception 类是指一些可以被捕获且可能恢复的异常情况,如数组下标越界、数字被零除产生异常、输入/输出异常等。

Exception 类又分为运行时异常和非运行时异常。RuntimeException 及其子类都属于运行时异常,如数字被零除产生异常、数组下标越界异常等。除了运行时异常以外的 Exception 子类都是非运行时异常。所有继承自 Error 和 RuntimeException 类的异常为未检查异常,其他的异常为已检查异常。Java 编译器要求 Java 程序必须处理已检查异常,而未检查异常、无法进行处理或是发生在程序运行时,编译器则没有硬性规定,建议开发者不处理未检查异常。

4.2 Java 异常处理机制

在介绍 Java 异常处理机制之前,先看一个例子。

```
public class ExcDemo01 {
    private String state = " 正常状态 ";
    public String getState() {
        return state;
    }
    public void setState(String state) {
        this.state = state;
    }
    public static void main(String[] args) {
        System.out.println("<-- 程序开始 -->");
        ExcDemo01 NullPointerCls = null;
        NullPointerCls.getState();
        System.out.println("<-- 程序结束 -->");
    }
}
```

不难发现,这段程序代码中 ExcDemo01 类的 NullPointerCls 对象没有实例化。这段程序代码的运行结果如下:

```
<-- 程序开始 -->
Exception in thread "main" java.lang.NullPointerException    at
example.code.exception.ExcDemo01.main(ExcDemo01.java:28)
```

从运行结果中可以得到以下信息:

(1)出现了 java.lang.NullPointerException,即通常说的空指针异常。

（2）在 ExcDemo01.java 文件产生 NullPointerCls.getState() 的异常代码。

（3）没有输出 <-- 程序结束 -->，说明程序中的异常没有得到处理从而导致异常终止。

上面的运行结果显然不是开发者所期望的。NullPointerException 是一种运行时异常、一种未检查异常，不用显式地进行异常处理。但是至少说明如果不对异常进行处理，程序就会异常终止。

Java 中有两种异常处理机制：捕获处理异常和声明抛出异常。与异常有关的关键字有 try、catch、throw、throws 和 finally。通过 try、catch、finally 关键字实现捕获处理异常，通过 throw、throws 关键字声明抛出异常。

4.2.1 捕获处理异常

捕获处理异常是先用 try 选定监控异常的范围，每个 try 代码块可以伴随一个或多个 catch 代码块，用于处理 try 代码块中的方法可能抛出的异常。catch 语句需要指明能够捕获处理的异常类型，这个类必须是 Throwable 的子类。捕获异常的顺序和 catch 语句的顺序有关，当捕获到一个异常时，剩下的 catch 语句就不再进行匹配。

在安排 catch 语句的顺序时，首先应该捕获最特殊的异常，然后逐渐一般化，也就是先安排子类，再安排父类。finally 语句保证在控制流转到程序的其他部分之前，finally 代码块中的程序都被执行。不管在 try 代码块中是否发生了异常事件，finally 代码块中的语句都会被执行。

 由于 finally 代码块中的语句总会被执行，因此要尽量避免在 finally 代码块中抛出异常。

下面是 Java 异常处理程序的基本形式：

```
try{
// 可能产生异常的代码
}
catch(异常类型 1 异常对象){
// 对异常类型 1 的处理
}
catch(异常类型 2 异常对象){
// 对异常类型 2 的处理
}
catch(异常类型 3 异常对象){
// 对异常类型 3 的处理
}
...      ...
...      ...
```

```
...    ...
catch(异常类型 n 异常对象){
// 对异常类型 n 的处理
}
finally{
// 总会执行的代码
}
```

下面是一个捕获处理异常的例子。

```
public class ExcDemo02 {
        private static void loadClass(String className)
            throws ClassNotFoundException {
            Class.forName(className);
        }
        public static void doSth(String className) {
            System.out.println("<-- 调用 loadClass() 方法开始 -->");
            try {
                System.out.println("<--try 的自述：在这里监控可能出现异常的代码 -->");
                loadClass(className);
            } catch (ClassNotFoundException e) {
                System.out.println("<--catch 的自述：出现异常在这里处理 -->");
            } finally {
                System.out.println("<--finally 的自述：不管是否出现异常，最后都要执行到这里 -->");
            }
            System.out.println("<-- 调用 loadClass() 方法结束 -->");
        }
    public static void main(String[] args) {
        System.out.println("<-- 调用 doSth() 方法开始 -->");
        doSth("java.lang.exception");
        System.out.println("<-- 调用 doSth() 方法结束 -->");
    }
}
```

这段程序代码的运行结果如下：

```
<-- 调用 doSth() 方法开始 -->
<-- 调用 loadClass() 方法开始 -->
<--try 的自述：在这里监控可能出现异常的代码 -->
<--catch 的自述：出现异常在这里处理 -->
<--finally 的自述：不管是否出现异常，最后都要执行到这里 -->
<-- 调用 loadClass() 方法结束 -->
```

<-- 调用doSth()方法结束 -->
```

从这段代码的运行结果中可以看出，doSth()方法捕获处理了loadClass()方法中出现的异常。

再看下面的例子。

```java
public class ExcDemo03 {
    private static void loadClass(String className)
            throws ClassNotFoundException {
        Class.forName(className);
    }
    public static void doSth(String className) {
        System.out.println("<-- 调用loadClass()方法开始 -->");
        try {
            loadClass(className);
        } catch (ClassNotFoundException e) {
System.out.println("<-- 这里只处理ClassNotFoundException-->");
        } catch (Exception e) {
            System.out.println("<- 这里处理所有的异常 ->");
        } finally {
        }
        System.out.println("<-- 调用loadClass()方法结束 -->");
    }
    public static void main(String[] args) {
        System.out.println("<-- 调用doSth()方法开始 -->");
        doSth("java.lang.exception");
        System.out.println("<-- 调用doSth()方法结束 -->");
    }
}
```

这段程序代码的运行结果如下：

```
<-- 调用doSth()方法开始 -->
<-- 调用loadClass()方法开始 -->
<-- 这里只处理ClassNotFoundException-->
<-- 调用loadClass()方法结束 -->
<-- 调用doSth()方法结束 -->
```

结果中为什么没有 <- 这里处理所有的异常 -> 呢？这个例子正好说明了前面提到的：捕获异常的顺序和catch语句的顺序有关。当捕获到一个异常时，剩下的catch语句就不再进行匹配。为了加深对这个规则的理解和认识，再来看下面的例子：

```java
public class ExcDemo04 {
```
```

```java
    private static void loadClass(String className)
            throws ClassNotFoundException {
        Class.forName(className);
    }
    public static void doSth(String className) {
        System.out.println("<-- 调用 loadClass() 方法开始 -->");
        try {
            loadClass(className);
        } catch (Exception e) {
            System.out.println("<- 这里处理所有的异常 ->");
        } catch (ClassNotFoundException e) {
System.out.println("<-- 这里只处理
ClassNotFoundException-->");
        } finally {
        }
        System.out.println("<-- 调用 loadClass() 方法结束 -->");
    }
    public static void main(String[] args) {
        System.out.println("<-- 调用 doSth() 方法开始 -->");
        doSth("java.lang.exception");
        System.out.println("<-- 调用 doSth() 方法结束 -->");
    }
}
```

这个例子和上面的例子有什么区别呢？只是把两个 catch 块的顺序交换了一下。这个例子正好解释了：在安排 catch 语句的顺序时，首先应该捕获最特殊的异常，然后逐渐一般化，也就是一般先安排子类，再安排父类。如果用 Eclipse 3.1 作为开发工具编写这段代码，不用编译 Eclipse 3.1 就会自动指出，第 2 个 catch 块是在任何情况下都不可能被执行的。

一个 try 语句可以在另一个 try 块内部，也就是说，try 块是可以被嵌套的。每次进入 try 块，try 都会按其前后关系入栈，如果内部 try 块中的异常不能找到相应的 catch 块进行处理，就从栈中弹出并检查下一个 try 块对应的 catch 块是否与之匹配，直到找到匹配的为止，或者由于没有与之匹配的 catch 块而交由 JVM 处理。当然，如果交由 JVM 处理，程序就会异常终止。

看下面的例子：

```java
public class ExcDemo05 {
    private static void loadClass(String className)
            throws ClassNotFoundException {
        Class.forName(className);
    }
    public static void doSth(String className) {
        try {
```

```
                try{
                        loadClass(className);
                }finally{
                }
        } catch (ClassNotFoundException e) {
                System.out.println("<-- 在这里处理内部的try块 -->");
        } finally {
        }
    }
    public static void main(String[] args) {
        System.out.println("<-- 调用doSth()方法开始 -->");
        doSth("java.lang.exception");
        System.out.println("<-- 调用doSth()方法结束 -->");
    }
}
```

这段程序代码的运行结果如下：

```
<-- 调用doSth()方法开始 -->
<-- 在这里处理内部的try块 -->
<-- 调用doSth()方法结束 -->
```

4.2.2 声明抛出异常

抛出异常也是生成异常对象的过程。异常或是由虚拟机生成，或是在程序中生成。在方法中，抛出异常对象是通过throw语句实现的。

如果方法中生成了一个异常，但是当这种方法不需要或不知道如何处理时，就应该声明抛出异常，使得异常对象可以从调用栈向上传播，直到有相应的方法捕获为止。因此，一旦出现异常就伴随着程序流程的跳转，但不建议用异常作为流程控制手段控制程序的流程。声明抛出异常是在方法声明中的throws子句中指明的。

下面是Java声明抛出异常程序的基本形式。

```
方法返回值类型 方法名(方法参数类型1 对象1,... …方法参数类型n 对象n) throws 异常类型1,… …异常类型n{
    ...     ...
    ...     ...
    ...     ...
    throw 异常类型1的对象
    ...     ...
    ...     ...
    ...     ...
```

```
...    ...
throw 异常类型 n 的对象
}
```

回到前面的一个例子：

```
public class ExcDemo02 {
   private static void loadClass(String className)
            throws ClassNotFoundException {
      Class.forName(className);
   }
   public static void doSth(String className) {
      System.out.println("<-- 调用 loadClass() 方法开始 -->");
      try {
System.out.println("<--try 的自述：在这里监控
可能出现异常的代码 -->");
         loadClass(className);
      } catch (ClassNotFoundException e) {
System.out.println("<--catch 的自述：
出现异常在这里处理 -->");
      } finally {
System.out.println("<--finally 的自述：
不管是否出现异常，最后都要执行到这里 -->");
      }
      System.out.println("<-- 调用 loadClass() 方法结束 -->");
   }
   public static void main(String[] args) {
      System.out.println("<-- 调用 doSth() 方法开始 -->");
      doSth("java.lang.exception");
      System.out.println("<-- 调用 doSth() 方法结束 -->");
   }
}
```

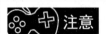注意　loadClass() 方法后面的 throws ClassNotFoundException，说明 loadClass() 方法没有捕获处理程序 Class.forName（className）所引发异常，loadClass() 方法使用 throws 关键字声明抛出了这个异常。调用 loadClass() 方法的 doSth() 方法使用 try-catch-finally 捕获处理了这个异常。而 doSth() 方法处理异常 ClassNotFoundException 的方式可以是将这个异常包装成更一般的 Exception，然后抛出。

这样做也是有意义的：可能 doSth() 方法不希望调用它的方法知道具体发生了什么的异常，也可能是调用 doSth() 方法的方法根本不关心究竟发生了什么异常，所有的异常都同样对待。

请看下面的例子：

```
public class ExcDemo05 {
    private static void loadClass(String className)
                throws ClassNotFoundException {
        Class.forName(className);
    }
    public static void doSth(String className) throws Exception {
        System.out.println("<-- 调用 loadClass() 方法开始 -->");
        try {
            loadClass(className);
        } catch (ClassNotFoundException e) {
            System.out.println("<-- 只向调用者声明可能出现 Exception 异常并不告知具体是什么异常 -->");
            throw new Exception();
        } finally {
        }
        System.out.println("<-- 调用 loadClass() 方法结束 -->");
    }
    public static void main(String[] args) {
        System.out.println("<-- 调用 doSth() 方法开始 -->");
        try {
            doSth("java.lang.exception");
        } catch (Exception e) {
            System.out.println("<-- 不管是什么异常，只要有异常就这样处理 -->");
        }
        System.out.println("<-- 调用 doSth() 方法结束 -->");
    }
}
```

这段程序代码的运行结果如下：

```
<-- 调用 doSth() 方法开始 -->
<-- 调用 loadClass() 方法开始 -->
<-- 只向调用者声明可能出现 Exception 异常并不告知具体是什么异常 -->
<-- 不管是什么异常，只要有异常就这样处理 -->
<-- 调用 doSth() 方法结束 -->
```

这段代码中，doSth() 方法使用 throw 关键字抛出了异常。

4.3 自定义异常

前面示例使用的都是 Java 类库中的异常。Java 还允许开发者定义自己的异常类，称为自定义异常。但需要遵守的规则是：自定义异常类必须是 Throwable 的子类，而更多情况下，自定义异常类都继承自 Exception 类。

> **提示**　更多情况下，自定义异常都继承自非运行时异常类。

下面的例子演示了如何编写自定义异常。

```java
import java.io.PrintStream;
import java.io.PrintWriter;
public class BuziException extends Exception {
    private static final long serialVersionUID = 1L;
    public synchronized Throwable fillInStackTrace() {
        System.out.println("<-- 添加自定义内容 -->");
        return super.fillInStackTrace();
    }
    public Throwable getCause() {
        System.out.println("<-- 添加自定义内容 -->");
        return super.getCause();
    }
    public String getLocalizedMessage() {
        System.out.println("<-- 添加自定义内容 -->");
        return super.getLocalizedMessage();
    }
    public String getMessage() {
        System.out.println("<-- 添加自定义内容 -->");
        return super.getMessage();
    }
    public StackTraceElement[] getStackTrace() {
        System.out.println("<-- 添加自定义内容 -->");
        return super.getStackTrace();
    }
    public synchronized Throwable initCause(Throwable arg0) {
        System.out.println("<-- 添加自定义内容 -->");
        return super.initCause(arg0);
    }
    public void printStackTrace() {
        System.out.println("<-- 添加自定义内容 -->");
        super.printStackTrace();
    }
    public void printStackTrace(PrintStream arg0) {
        System.out.println("<-- 添加自定义内容 -->");
        super.printStackTrace(arg0);
    }
    public void printStackTrace(PrintWriter arg0) {
        System.out.println("<-- 添加自定义内容 -->");
```

```
            super.printStackTrace(arg0);
        }
        public void setStackTrace(StackTraceElement[] arg0) {
            System.out.println("<-- 添加自定义内容 -->");
            super.setStackTrace(arg0);
        }
        public String toString() {
            System.out.println("<-- 添加自定义内容 -->");
            return super.toString();
        }
}
```

上面示例定义了名为 BuziException 的异常类，继承自 Exception，覆盖了 Throwable 类所有的方法。这里只是一个示例，开发者可以根据实际情况有选择地覆盖方法，添加有意义的自定义内容。

4.4 自定义异常的综合应用

下面通过一个例子来加深对异常处理机制和自定义异常的理解和认识。

首先自定义名为 BaseException 的异常。

```
import java.io.PrintStream;
import java.io.PrintWriter;
public class BaseException extends Exception {
    private static final long serialVersionUID = 1L;
    public BaseException(String msg){
        super(msg);
    }
    public synchronized Throwable fillInStackTrace() {
        System.out.println("<-- 执行 fillInStackTrace() 方法 -->");
        return super.fillInStackTrace();
    }
    public Throwable getCause() {
        System.out.println("<-- 执行 getCause() 方法 -->");
        return super.getCause();
    }
    public String getLocalizedMessage() {
        System.out.println("<-- 执行 getLocalizedMessage() 方法 -->");
        return super.getLocalizedMessage();
    }
    public String getMessage() {
        System.out.println("<-- 执行 getMessage() 方法 -->");
```

```
            return super.getMessage();
    }
    public StackTraceElement[] getStackTrace() {
            System.out.println("<-- 执行getStackTrace()方法 -->");
            return super.getStackTrace();
    }
    public synchronized Throwable initCause(Throwable arg0) {
            System.out.println("<-- 执行initCause(Throwable arg0)
方法 -->");
            return super.initCause(arg0);
    }
    public void printStackTrace() {
            System.out.println("<-- 执行printStackTrace()方法 -->");
            super.printStackTrace();
    }
    public void printStackTrace(PrintStream arg0) {
System.out.println("<-- 执行printStackTrace(PrintStream arg0)方法 -->");
            super.printStackTrace(arg0);
    }
    public void printStackTrace(PrintWriter arg0) {
            System.out.println("<-- 执行printStackTrace
(PrintWriter arg0)方法 -->");
            super.printStackTrace(arg0);
    }
    public void setStackTrace(StackTraceElement[] arg0) {
            System.out.println("<-- 执行setStackTrace
(StackTraceElement[] arg0)方法 -->");
            super.setStackTrace(arg0);
    }
    public String toString() {
            System.out.println("<-- 执行toString()方法 -->");
            return super.toString();
    }
}
```

构造抛出 BaseException 的方法，分别应用两种异常处理机制处理抛出的 BaseException。

```
public class ExcDemo06 {
    private static void excptionSource() throws BaseException{
            System.out.println("<--excptionSource()方法生来就是
为了抛出异常-->");
            throw new BaseException("这是BaseException");
    }
```

```java
    public static void catchException(){
        try {
            excptionSource();
        } catch (BaseException e) {
            System.out.println("<-- 异常产生时的附加信息是: "+e.getMessage()+"-->");
        }
    }
    public static void throwException() throws BaseException{
        excptionSource();
    }
    public static void main(String[] args) {
        System.out.println("<-- 调用 catchException() 方法开始 -->");
        catchException();
        System.out.println("<-- 调用 catchException() 方法结束 -->");
        try {
            System.out.println("<-- 调用 throwException() 方法开始 -->");
            throwException();
            System.out.println("<-- 调用 throwException() 方法结束 -->");
        } catch (BaseException e) {
            System.out.println("<-- 既然 throwException() 方法不处理异常, 就由 main() 方法来处理, 不然程序就异常中断了 -->");
            System.out.println("<-- 异常产生时的附加信息是: "+e.getMessage()+"-->");
        }
    }
}
```

这段程序代码的运行结果如下：

```
<-- 调用 catchException() 方法开始 -->
<--excptionSource() 方法生来就是为了抛出异常 -->
<-- 执行 fillInStackTrace() 方法 -->
<-- 执行 getMessage() 方法 -->
<-- 异常产生时的附加信息是：这是 BaseException-->
<-- 调用 catchException() 方法结束 -->
<-- 调用 throwException() 方法开始 -->
<--excptionSource() 方法生来就是为了抛出异常 -->
<-- 执行 fillInStackTrace() 方法 -->
<-- 既然 throwException() 方法不处理异常, 就由 main() 方法来处理, 不然程序就异常中断了 -->
<-- 执行 getMessage() 方法 -->
```

```
<-- 异常产生时的附加信息是：这是 BaseException -->
```

从上面的运行结果中可以看出：

（1）在抛出 BaseException 的同时，会调用 BaseException 的 fillInStackTrace() 方法将异常放进栈中。

（2）可以通过调用 BaseException 对象的 getMessage() 方法获得生成 BaseException 对象时的附加信息。

（3）由于 throwException() 方法中的 excptionSource(); 语句出现异常，异常被抛出，因此不会执行 System.out.println（"<-- 调用 throwException() 方法结束 -->"）; 语句。

4.5 实例练习：异常的综合应用

在本章的最后，我们将综合运用前面学习的知识给出一个完整的实例。

该实例综合运用两种异常处理机制，分别对 Java 类库中的异常和自定义异常进行处理，其中自定义异常类使用前面例子中的 BuziException。实例的源代码如下：

```java
import java.sql.SQLException;
public class DaoClass {
    private void prepare() throws SQLException {
        System.out.println("<-- 构造一个 SQLException 异常 -->");
        throw new SQLException();
    }
    private void todo() {
        System.out.println("<-- 一个没有异常的方法 -->");
    }
    public void callPrepare() throws SQLException {
        System.out.println("<-- 调用 prepare 方法，对其抛出的异常不做捕获处理直接抛出 -->");
        prepare();
        todo();
    }
}
import java.sql.SQLException;
public class BuziClass {
    public void CallDaoMethod() throws BuziException {
        DaoClass dao = new DaoClass();
        try {
            dao.callPrepare();
        } catch (SQLException e) {
            System.out.println("<-- 捕获 callPrepare 方法抛出的异常,
```

```
        将其封装成自定义异常BuziException抛出 -->");
                throw new BuziException();
        }
    }
    public static void main(String[] args) {
        BuziClass buzi = new BuziClass();
        try {
            buzi.CallDaoMethod();
        } catch (BuziException e) {
            System.out.println("<-- 调用CallDaoMethod方法，捕获其抛出的异常 -->");
            e.printStackTrace();
        }
    }
}
```

4.6 要点总结

本章首先对异常的概念和没有异常概念的程序错误处理进行了简要介绍；然后介详述了Java中异常类的继承结构和分类；重点讲解了Java的两种异常处理机制：捕获处理异常和声明抛出异常，同时介绍了自定义异常的方法；最后通过一个实例讲解了从构造自定义异常到异常处理的过程。

4.7 编程练习

1. 填空题

（1）Java中，异常分为 _____ 和 _____ 两类。

（2）异常是在运行时程序代码序列中产生的一种 _____、_____。

（3）在Java中，每个异常都是一个 _____，它是 _____ 类或其子类的实例。

（4）Throwable类有两个重要的子类：_____ 和 _____。

（5）Exception又分为 _____ 异常和 _____ 异常。

2. 选择题

（1）下列哪种方法不是Throwable类的构造方法 _____。

  A. Throwable()　　　　　　　　　　　　B. Throwable(String message)

  C. Throwable(Throwable cause, String message)　　D. Throwable(Throwable cause)

（2）下列异常类中不是继承自 RuntimeException 类的是 _____。

    A. ArithmticException   B. NullPointerException

    C. IndexOutOfBoundException   D. FileNotFoundException

（3）下列不是用来捕获处理异常的关键字是 _____。

    A. throws   B. try   C. catch   D. finally

（4）下列关键字哪个是用来在方法名后声明抛出异常的 _____。

    A. throw   B. throws   C. enum   D. final

（5）在抛出异常时，会自动调用这个异常的是哪个方法 _____。

    A. clone()   B. getMessage()   C. fillInStackTrace()   D. toString()

3. 问答题

（1）简述在没有异常机制的程序语言中如何进行错误识别和处理。

（2）简述 Java 异常处理机制（两种方式）。

4. 编程练习

参照本章中的示例编写一个自定义异常类，要求继承 Exception 类，在 main 方法中抛出这个异常并进行捕获处理。

第 5 章
Java 线程

Java 是少数的几种支持多线程的语言之一。大多数程序语言只能运行单独的程序块，无法同时运行多个不同的程序块，而 Java 的多线程机制就弥补了这个缺憾，可以让不同的程序块同时运行，这样程序运行会更为顺畅，性能也更高，同时达到多任务处理的目的。到目前为止，本书所介绍过的实例都是单线程程序，也就是说，执行的 Java 程序只做一件事。例如，一个工匠做一把椅子和多个工匠同时做一把椅子，就是单线程和多线程的区别，显然多线程机制有更高的性能。本章主要介绍多线程的概念及有关多线程的编程方法。

5.1 多线程及线程简介

多线程使程序可以同时存在多个执行片段，根据不同的条件和环境同步或异步工作。线程与进程的实现原理类似，但它们的服务对象不同，进程代表操作系统平台中运行的一个程序，而一个程序中可能包含多个线程。

进程是程序的一次动态执行过程，需要经历代码加载、代码执行到代码执行完成的一个完整过程，这个过程也是进程本身从产生、就绪、执行、阻塞、再执行到最终消亡的过程。

所谓的多线程，是指一个进程在执行过程中可以产生多个更小的程序单元，这些更小的程序单元称为线程，这些线程同时存在、同时运行。进程与线程的区别如图 5-1 所示。

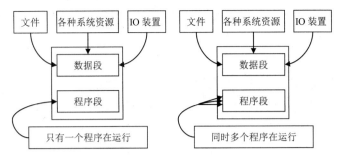

图 5-1 线程与进程的区别

5.2 线程的创建

在 Java 语言中，线程也是一种对象，但并非任何对象都可以成为线程，只有实现 Runnable 接口或继承了 Thread 类的对象才行。

1. 线程的创建方式

Java 的线程是通过 java.lang.Thread 类实现的。当生成一个 Thread 类的对象之后，一个新的线程就产生了。线程实例表示 Java 解释器中真正的线程，通过它可以启动线程、终止线程、线程挂起等。每个线程都是通过某个特定 Thread 对象的方法 run() 来完成其操作的，方法 run() 称为线程体。先来了解一下 Thread 类。

Thread 类的 8 个构造函数如下：

```
Thread ()
Thread (Runnable target)
Thread (Runnable target, String name)
Thread (String name)
Thread (ThreadGroup group, Runnable target)
Thread (ThreadGroup group, Runnable target, String name)
Thread (ThreadGroup group, Runnable target, String name, long stackSize)
Thread (ThreadGroup group, String name)
```

Thread 类的常用方法如下：

```
static Thread currentThread()
static int activeCount()
long getId()
String getName ()
int getPriority()
Thread.State getState ()
ThreadGroup getThreadGroup ()
void interrupt()
```

```
static boolean interrupted()
boolean isAlive()
boolean isDaemon()
boolean isInterrupted()
void join()
void join(long millis)
void join(long millis, int nanos)
void run()
void setDaemon(boolean on)
void setName(String name)
void setPriority(int newPriority)
static void sleep(long millis)
static void sleep(long millis, int nanos)
void start()
static void yield()
```

实现 Runnable 接口的类都可以充当线程体，在接口 Runnable 中只定义了一种方法。

```
void run()
```

任何实现接口 Runnable 的对象都可以作为一个线程的目标对象，类 Thread 本身也实现了接口 Runnable。

2. 继承 Thread 类

定义一个线程类，它继承线程类 Thread 并重写其中的 run() 方法，以实现用户所需的功能，实例化自定义的 Thread 类，使用 start() 方法启动线程。由于 Java 只支持单重继承，因此用这种方法定义的类不能再继承其他父类。

下面的示例定义了一个继承自 Thread 类的 SimpleThread 类，该类将创建的两个线程同时在控制台输出信息，从而实现两个任务输出信息的交叉显示。

```java
public class SimpleThread extends Thread{
    /**
     * @param name
     * 参数 name 为线程名称
     */
    public SimpleThread(String name){
        setName(name);
    }
    @Override
    public void run() {                              // 覆盖 run() 方法
        int i = 0;
        while (i++ < 5) {
            try {
                System.out.println(getName() + "执行步骤" + i);
```

```
                    Thread.sleep(1000);                    // 休眠1秒
                } catch (Exception e) {
                    e.printStackTrace();
                }
            }
        }
        public static void main(String[] args) {
            // 创建线程1
            SimpleThread st1 = new SimpleThread(" 线程1");
            // 创建线程2
            SimpleThread st2 = new SimpleThread("======线程2");
            // 启动线程1
            st1.start();
            // 启动线程2
            st2.start();
        }
    }
```

这段程序代码的运行结果如下：

```
======线程2执行步骤1
线程1执行步骤1
======线程2执行步骤2
线程1执行步骤2
线程1执行步骤3
======线程2执行步骤3
======线程2执行步骤4
线程1执行步骤4
线程1执行步骤5
======线程2执行步骤5
```

从程序的执行结果中可以发现，因为现在的两个进程对象是交错运行的，哪个线程对象抢到 CPU 资源，哪个线程就可以运行，所以程序的执行结果并不是固定的。在线程启动时虽然调用的是 start（）方法，但实际调用的却是 run() 方法定义的主体。

 启动线程不能直接使用 run() 方法，因为线程的运行需要本机操作系统的支持。如果一个类通过继承 Thread 类实现，就只能调用一次 start() 方法；如果调用多次，就会抛出 Exception in thread "main" java.lang.IllegalThreadStateException。

3. 继承 Runnable 接口

Runnable 是 Java 语言中用以实现线程的接口，任何实现线程功能的类都必须实现这个接口。Thread 类就是因为实现了 Runnable 接口，所以继承的类才具有相应的线程功能。

Runnable 接口中定义了一个 run() 方法，在实例化 Thread 对象时，可以传入实现 Runnable 接口的对象作为参数，Thread 类会调用 Runnable 对象的 run() 方法，继而执行 run() 方法中的内容。

下面的示例定义了一个实现接口 Runnable 的 SimpleRunnable 类作为线程的目标对象。该类在 run() 方法中每间隔 0.5 秒，在控制台输出一个"@"，直到输出 15 个"@"字符。

```
public class SimpleRunnable implements Runnable {
    // 覆盖 run() 方法
    @Override
    public void run() {
        int i = 15;
        while (i-- >=1) {
            try {
                System.out.print("@ ");
                Thread.sleep(500);
            } catch (Exception e) {
                e.printStackTrace();
            }
        }
    }
    public static void main(String[] args) {
        // 创建线程 A
        Thread t = new Thread(new SimpleRunnable(),"线程A");
        // 启动线程 A
        t.start();
    }
}
```

这段程序代码的运行结果如下：

@ @ @ @ @ @ @ @ @ @ @ @ @ @ @

显而易见，这段程序代码是上一个示例的另外一种实现方式。

不管是线程体的哪种实现方式，从结果中都不难发现三个线程是并行的。通常情况下，因为计算机只有一个 CPU，所以在某个时刻只能有一个线程在运行，线程的并发调度执行在 Java 语言设计之初就已经被设计者考虑在内了。

最后比较一下实现线程体的两种方式：使用直接继承 Thread 类的方式比较简单，但由于 Java 的单继承机制，其局限不言而喻。使用 Runnable 接口首先不存在前一种方法的局限，而且这种方式是将虚拟 CPU（Thread）及其子类与代码和数据分开，符合线程的概念模型。开发者可以根据不同的需求结合这两种方式的特点选择合适的线程体实现方式。

 掌握实现线程体的方式是 Java 线程的基础，同时也要熟悉两种实现线程体方式的适用范围。

5.3 线程的状态

线程有 4 种状态：创建状态、可运行状态、不可运行状态和死亡状态。

1. 创建状态

在实例化一个线程对象后，线程就处于创建状态。处于创建状态的线程，系统不为其分配资源。

2. 可运行状态

当线程对象调用了 run() 方法后，线程就处于可运行状态。处于可运行状态的线程，系统为其分配了所需的系统资源。这里要注意区分可运行和运行。线程对象调用了 run() 方法后，只表明线程处于可运行状态，不代表正在运行。通常情况下，因为计算机只有一个 CPU，所以同一时刻只能运行处于可运行状态线程中的一个，通过 Java 运行系统的调度达到共享 CPU 的目的。同一时间处于可运行状态的线程可能有很多个，而正在运行的线程却只有一个。

3. 不可运行状态

不可运行状态也称为阻塞状态，由于某种原因系统不能执行线程的状态。在这种情况下，即使 CPU 处于空闲状态，线程也不能被执行。

线程成为不可运行状态可能是因为调用了 sleep() 方法、suspend() 方法、wait() 方法或输入输出 / 流中发生线程阻塞。

4. 死亡状态

线程执行完或是调用了 stop() 方法都会进入死亡状态。

如图 5-2 所示为 Java 线程的不同状态及状态之间转换所调用的方法。

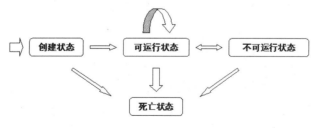

图 5-2 Java 线程状态

5.4 线程的调度

Java 提供了一个线程调度器来调度程序启动后进入可运行状态的所有线程。线程调度器按照线程的优先级决定处于可运行状态线程的运行顺序。线程调度器按线程的优先级高低选择优先级高的线程先运行。

线程调度是抢先式调度，即如果在当前线程运行过程中，一个更高优先级的线程进入可运行状态，这个线程就会立即被调度运行。

抢先式调度又分为时间片方式和独占方式。

在时间片方式下，当前活动线程运行完当前时间片后，如果有其他处于就绪状态的相同优先级线程，系统就会将运行权交给其他就绪状态的同优先级线程；当前活动线程转入等待运行队列，等待下一个时间片的调度。

在独占方式下，当前活动线程一旦获得运行权，将一直执行下去，直到执行完毕或由于某种原因主动放弃 CPU，或者有高优先级的线程处于可运行状态。

线程主动放弃 CPU 可能是如下原因：

（1）线程调用了 yield() 或 sleep() 方法主动放弃。

（2）线程调用 wait() 方法。

（3）由于当前线程进行 I/O 访问、外存读写、等待用户输入等操作，因此导致线程阻塞。

5.5 线程的优先级

线程有 10 个优先级，由低到高分别用 1~10 表示，默认值为 5。Thread 类中定义了三个静态变量，如下：

```
Thread.MIN_PRIORITY 代表最低优先级 1
Thread.NORM_PRIORITY 代表默认优先级 5
Thread.MAX_PRIORITY 代表最高优先级 10
```

除此之外，还可以通过 Thread 类的两个方法对优先级进行操作。

```
public final void setPriority(int newPriority)
public final int getPriority()
```

下面通过代码来演示三种不同优先级的线程执行结果。

```java
public class PriorityDemo01 implements Runnable {   // 实现 Runnable 接口
    @Override
    public void run() {                              // 覆写 run() 方法
        for (int i = 0; i < 5; i++) {                // 循环 5 次
            try {
                Thread.sleep(500);                   // 线程休眠
            } catch (Exception e) {                  // 需要异常处理
```

```java
                        e.printStackTrace();
                }
                System.out.println(Thread.currentThread().getName() +
"运行, i=" + i);                              // 输出线程名称
        }
    }
    public static void main(String[] args) {
            // 实例化线程对象
Thread t1 = new Thread(new PriorityDemo01(),"线程1");
            // 实例化线程对象
Thread t2 = new Thread(new PriorityDemo01(),"线程2");
            // 实例化线程对象
Thread t3 = new Thread(new PriorityDemo01(),"线程3");
            // 设置线程优先级为最低
t1.setPriority(Thread.MIN_PRIORITY);
            // 设置线程优先级为最高
t2.setPriority(Thread.MAX_PRIORITY);
            // 设置线程优先级为默认
t3.setPriority(Thread.NORM_PRIORITY);
            t1.start();                      // 启动线程
            t2.start();                      // 启动线程
            t3.start();                      // 启动线程
    }
}
```

这段程序代码的运行结果如下：

```
线程2 运行, i=0
线程3 运行, i=0
线程1 运行, i=0
线程2 运行, i=1
线程3 运行, i=1
线程1 运行, i=1
线程2 运行, i=2
线程3 运行, i=2
线程1 运行, i=2
线程2 运行, i=3
线程3 运行, i=3
线程1 运行, i=3
线程2 运行, i=4
线程3 运行, i=4
线程1 运行, i=4
```

从程序的运行结果可以看到，线程将根据其优先级的大小来决定哪个线程会先运行，但

是读者一定要注意的是，并非线程的优先级越高就一定会先运行，哪个线程先运行由 CPU 的调度决定。

Java 程序启动运行时，JVM 会自动创建一个线程，称为主线程。

主线程的重要性和特殊性表现在以下两个方面。

（1）是产生其他线程的线程。

（2）通常执行各种关闭操作，是最后结束的。

虽然主线程有些特殊，无须手动创建，但还是一个线程类的对象。可以通过 Thread 类的方法获得主线程的引用，从而达到操作主线程的目的。

下面的程序演示了主方法 main 的优先级是 NORM_PRIORITY。

```java
public class GetMainThread {
    public static void main(String[] args) {
        Thread thread = Thread.currentThread();
        System.out.println("<-- 当前主线程的 ID 是 "+thread.getId()+"-->");
        System.out.println("<-- 当前主线程的名称是 "+thread.getName()+"-->");
        System.out.println("<-- 当前主线程的优先级是 "+thread.getPriority()+"-->");
        System.out.println("<-- 当前主线程所在线程组是 "+thread.getThreadGroup().getName()+"-->");
    }
}
```

这段程序代码的运行结果如下：

```
<-- 当前主线程的 ID 是 1-->
<-- 当前主线程的名称是 main-->
<-- 当前主线程的优先级是 5-->
<-- 当前主线程所在线程组是 main-->
```

下面的程序演示了线程的礼让，即使用 yield() 方法将一个线程的操作暂时让给其他线程执行。代码如下：

```java
public class PriorityDemo02 implements Runnable{
    @Override
    public void run() {                              // 覆写 run() 方法
        for (int i = 0; i < 5; i++) {                // 循环 5 次
            System.out.println(Thread.currentThread().getName() + " 运行, i=" + i);    // 输出线程名称
```

```
                if (i == 3) {
                    System.out.println("线程礼让: ");
                    Thread.currentThread().yield();     // 线程礼让
                }
            }
        }
        public static void main(String[] args) {
            // 实例化 PriorityDemo02 对象
PriorityDemo02 my = new PriorityDemo02();
            Thread t1 = new Thread(my,"线程 1");        // 定义线程对象
            Thread t2 = new Thread(my,"线程 2");        // 定义线程对象
            t1.start();                                 // 启动线程
            t2.start();                                 // 启动线程
        }
}
```

这段程序代码的运行结果如下:

```
线程 1 运行, i=0
线程 2 运行, i=0
线程 1 运行, i=1
线程 2 运行, i=1
线程 2 运行, i=2
线程 2 运行, i=3
线程礼让:
线程 1 运行, i=2
线程 1 运行, i=3
线程 2 运行, i=4
线程礼让:
线程 1 运行, i=4
```

从程序的运行结果中可以发现，每当线程满足条件（i= =3）时，就会将本线程暂停，而让其他线程先运行。

5.6 守护线程

线程默认都是非守护线程，非守护线程也称作用户线程。当所有用户线程都结束时，守护线程也立即结束。因为守护线程随时会结束，所以守护线程所做的工作应该是可以随时结束而不影响运行结果的工作。

 提示　守护线程在 Java 线程应用开发中经常使用。

通过调用 Thread 类的方法将线程设置为守护线程。

```
public final void setDaemon(boolean on)
```

为了更好地理解守护线程的概念,下面是使用守护线程的示例。

首先构造一个模拟播放音乐的线程类。

```java
package example.code.thread;
public class MusicThread extends Thread {
   @Override
   public void run() {
        while (true) {
             System.out.println("<-- 音乐播放中......-->");
             try {
                  sleep(100);
             } catch (InterruptedException e) {
                  e.printStackTrace();
             }
        }
   }
}
```

然后构造一个模拟安装程序的线程类。

```java
public class InstallThread extends Thread {
   @Override
   public void run() {
        System.out.println("<-- 安装开始 -->");
        for (int i = 0; i <= 100; i = i + 10) {
             System.out.println("<-- 已安装 " + i + "%-->");
             try {
                  sleep(100);
             } catch (InterruptedException e) {
                  e.printStackTrace();
             }.
        }
        System.out.println("<-- 安装结束 -->");
   }
   public static void main(String[] args) {
        MusicThread music = new MusicThread();
        music.setDaemon(true);
        music.start();
        InstallThread install = new InstallThread();
        install.start();
```

```
        }
    }
```

这段程序代码的运行结果如下：

```
<-- 在这里定义 MusicThread 类的构造方法 -->
<-- 在这里定义 InstallThread 类的构造方法 -->
<-- 音乐播放中......-->
<-- 安装开始 -->
<-- 已安装 0%-->
<-- 音乐播放中......-->
<-- 已安装 10%-->
<-- 音乐播放中......-->
<-- 已安装 20%-->
<-- 音乐播放中......-->
<-- 已安装 30%-->
<-- 音乐播放中......-->
<-- 已安装 40%-->
<-- 音乐播放中......-->
<-- 已安装 50%-->
<-- 音乐播放中......-->
<-- 已安装 60%-->
<-- 音乐播放中......-->
<-- 已安装 70%-->
<-- 音乐播放中......-->
<-- 已安装 80%-->
<-- 音乐播放中......-->
<-- 已安装 90%-->
<-- 音乐播放中......-->
<-- 已安装 100%-->
<-- 音乐播放中......-->
<-- 安装结束 -->
```

上面的示例中，若模拟播放音乐的线程单独运行，则不会停止。但作为模拟安装程序的守护线程，一旦作为用户线程的模拟安装线程结束，模拟播放音乐的线程也会立即结束。

5.7 线程同步

当多个线程共享同一个变量等资源时，需要确保资源在某一时刻只有一个线程占用，这个过程就是线程同步。信号量是同步中的一个重要概念。信号量是一个对象，也是互斥体。当一个线程进入互斥体，互斥体就会锁定，此时任何试图进入互斥体的线程都必须等待这个线程出来。

 Java 线程同步本身的复杂性决定了要正确处理线程同步不是一件容易的事情。限于本书的涉众和篇幅，这里只介绍线程同步的基本概念和处理方法。如果读者有兴趣，可以查找相关资料进行专项学习。

为了深刻理解线程同步的概念，先来看一个没有使用线程同步的例子。

首先构造一个电话类，定义一个打电话的方法。

```java
public class PhoneCall {
    public static void call(String name) {
        try {
            System.out.println("<--" + name + " 拨打电话 -->");
            Thread.sleep(100);
            System.out.println("<--" + name + " 正在通话中......-->");
            Thread.sleep(100);
            System.out.println("<--" + name + " 挂断电话 -->");
        } catch (InterruptedException e) {
            e.printStackTrace();
        }
    }
}
```

然后构造一个调用电话类打电话方法的线程类。看看一部电话在没有同步情况下的工作状况。

```java
public class Call extends Thread {
    public Call(String arg0) {
        super(arg0);
        System.out.println("<-- 在这里定义 Call 类的构造方法 -->");
    }
    @Override
    public void run() {
        PhoneCall.call(getName());
    }
    public static void main(String[] args) {
        Call first = new Call("First");
        Call second = new Call("Second");
        Call third = new Call("Third");
        first.start();
        second.start();
        third.start();
    }
}
```

这段程序代码的运行结果如下：

```
<-- 在这里定义 Call 类的构造方法 -->
<-- 在这里定义 Call 类的构造方法 -->
<-- 在这里定义 Call 类的构造方法 -->
<--First 拨打电话 -->
<--Second 拨打电话 -->
<--Third 拨打电话 -->
<--First 正在通话中......-->
<--Second 正在通话中......-->
<--Third 正在通话中......-->
<--First 挂断电话 -->
<--Second 挂断电话 -->
<--Third 挂断电话 -->
```

从运行结果中不难看出，在 First 拨打电话的等待过程中，Second 也拨打了电话。Second 进入等待时，Third 开始拨打电话。这显然是不应该发生的，所以我们需要同步。

同样是这个情景，再看看使用同步的例子。

首先构造一个电话类，定义一个打电话的方法。

```java
public class SynPhoneCall {
    public synchronized static void call(String name) {
        try {
            System.out.println("<--" + name + " 拨打电话 -->");
            Thread.sleep(100);
            System.out.println("<--" + name + " 正在通话中......-->");
            Thread.sleep(100);
            System.out.println("<--" + name + " 挂断电话 -->");
        } catch (InterruptedException e) {
            e.printStackTrace();
        }
    }
}
```

然后构造一个调用电话类打电话方法的线程类。

```java
public class SynCall extends Thread{
    public SynCall() {
        super();
        System.out.println("<-- 在这里定义 SynCall 类的构造方法 -->");
    }
    public SynCall(Runnable arg0, String arg1) {
        super(arg0, arg1);
        System.out.println("<-- 在这里定义 SynCall 类的构造方法 -->");
    }
    public SynCall(Runnable arg0) {
```

```java
        super(arg0);
        System.out.println("<-- 在这里定义 SynCall 类的构造方法 -->");
    }
    public SynCall(String arg0) {
        super(arg0);
        System.out.println("<-- 在这里定义 SynCall 类的构造方法 -->");
    }
    public SynCall(ThreadGroup arg0, Runnable arg1, String arg2, long arg3) {
        super(arg0, arg1, arg2, arg3);
        System.out.println("<-- 在这里定义 SynCall 类的构造方法 -->");
    }
public SynCall(ThreadGroup arg0, Runnable arg1, String arg2) {
        super(arg0, arg1, arg2);
        System.out.println("<-- 在这里定义 SynCall 类的构造方法 -->");
    }
    public SynCall(ThreadGroup arg0, Runnable arg1) {
        super(arg0, arg1);
        System.out.println("<-- 在这里定义 SynCall 类的构造方法 -->");
    }
    public SynCall(ThreadGroup arg0, String arg1) {
        super(arg0, arg1);
        System.out.println("<-- 在这里定义 SynCall 类的构造方法 -->");
    }
    @Override
    public void run() {
        SynPhoneCall.call(getName());
    }
    public static void main(String[] args) {
        SynCall first = new SynCall("First");
        SynCall second = new SynCall("Second");
        SynCall third = new SynCall("Third");
        first.start();
        second.start();
        third.start();
    }
}
```

这段程序代码的运行结果如下：

```
<-- 在这里定义 SynCall 类的构造方法 -->
<-- 在这里定义 SynCall 类的构造方法 -->
<-- 在这里定义 SynCall 类的构造方法 -->
<--First 拨打电话 -->
<--First 正在通话中......-->
```

```
<--First 挂断电话 -->
<--Second 拨打电话 -->
<--Second 正在通话中......-->
<--Second 挂断电话 -->
<--Third 拨打电话 -->
<--Third 正在通话中......-->
<--Third 挂断电话 -->
```

仔细观察这个示例的代码与上一个示例的代码，本质上的差别就在于 synchronized 关键字。除了这种同步的实现方式外，还有一种实现同步的方式，也需要使用 synchronized 关键字。

再看下面的例子。

同样的情景，首先构造一个电话类，定义一个打电话的方法。

```java
public class PhoneCalls {
    private String phoneName = "";
    public PhoneCalls(String name) {
        this.phoneName = name;
    }
    public void call(String name) {
        try {
            System.out.println("<--" + name + " 拨打 "+this.phoneName+" 电话 -->");
            Thread.sleep(100);
            System.out.println("<--" + name + " 正在通话中......-->");
            Thread.sleep(100);
            System.out.println("<--" + name + " 挂断 "+this.phoneName+" 电话 -->");
        } catch (InterruptedException e) {
            e.printStackTrace();
        }
    }
}
```

然后构造一个调用电话类打电话方法的线程类。

```java
public class SynCalls implements Runnable {
    private String name = "";
    private PhoneCalls phone = null;
    private Thread thread = null;
    public SynCalls(String name, PhoneCalls phone) {
        this.name = name;
        this.phone = phone;
        this.thread = new Thread(this);
    }
```

```java
    public void start(){
        thread.start();
    }
    public void run() {
        synchronized(this.phone){
            this.phone.call(this.name);
        }
    }
    public static void main(String[] args) {
        PhoneCalls phone = new PhoneCalls("营部");
        SynCalls first = new SynCalls("First",phone);
        SynCalls second = new SynCalls("Second",phone);
        SynCalls third = new SynCalls("Third",phone);
        first.start();
        second.start();
        third.start();
    }
}
```

这段程序代码的运行结果如下：

```
<--First 拨打营部电话 -->
<--First 正在通话中 ......-->
<--First 挂断营部电话 -->
<--Second 拨打营部电话 -->
<--Second 正在通话中 ......-->
<--Second 挂断营部电话 -->
<--Third 拨打营部电话 -->
<--Third 正在通话中 ......-->
<--Third 挂断营部电话 -->
```

正如上例中所示，使用下面的方式，同样可以实现同步。

```
synchronized(mutex){

}
```

其中 mutex 就是互斥体，是一个对象。对于上面的例子，mutex 是一个电话实例，即营部电话。

5.8 实例练习：线程综合应用

在本章的最后，我们将综合运用前面所学的知识给出一个完整的实例。

该实例综合运用了两种实现线程体的方式、守护线程和主线程等知识。实例的源代码如下：

首先，通过实现 Runnable 接口实现线程体。

```java
public class ThreadUseRunnable implements Runnable {
    private String name;
    public ThreadUseRunnable(String name) {
        this.name = name;
    }
    public String getName() {
        return name;
    }
    public void run() {
        System.out.println("<--" + this.getName() + " 执行开始 -->");
        for (int i = 0; i < 10; i++) {
            System.out.println("<--" + this.getName() + " 执行步骤 " + i + "-->");
            try {
                Thread.sleep((int) (Math.random() * 1000));
            } catch (InterruptedException e) {
                e.printStackTrace();
            }
        }
        System.out.println("<--" + this.getName() + " 执行结束 -->");
    }
}
```

再通过继承 Thread 类的方式实现线程体。

```java
package example.code.thread;
public class ThreadUseExtends extends Thread {
    public ThreadUseExtends(String arg0) {
        super(arg0);
    }
    public void run() {
        System.out.println("<--" + this.getName() + " 执行开始 -->");
        for (int i = 0; i < 10; i++) {
            System.out.println("<--" + this.getName() + " 执行步骤 " + i + "-->");
            try {
                sleep((int) (Math.random() * 1000));
```

```
                } catch (InterruptedException e) {
                    e.printStackTrace();
                }
            }
            System.out.println("<--" + this.getName() + " 执行结束 -->");
        }
    }
```

然后通过继承 Thread 类的方式实现守护线程。

```
package example.code.thread;
public class DaemonThread extends Thread {
    public DaemonThread(String arg0) {
        super(arg0);
    }
    @Override
    public void run() {
        while (true) {
            System.out.println("<--" + this.getName() + " 执行中...-->");
            try {
                sleep(300);
            } catch (InterruptedException e) {
                e.printStackTrace();
            }
        }
    }
}
```

最后综合使用这三个线程和主线程。

```
package example.code.thread;
public class MultiThread {
    public static void main(String[] args) {
        System.out.println("<-- 执行开始 -->");
        Thread thread1 = new ThreadUseExtends("线程1");
        Thread thread2 = new Thread(new ThreadUseRunnable("线程2"));
        Thread thread3 = new DaemonThread("守护线程");
        thread3.setDaemon(true);
        thread3.start();
        thread1.start();
        thread2.start();
        while (thread1.isAlive() || thread2.isAlive()) {
```

```
                try {
                        Thread.sleep(100);
                } catch (InterruptedException e) {
                        e.printStackTrace();
                }
        }
                System.out.println("<-- 执行结束 -->");
    }
}
```

5.9 要点总结

本章首先简单介绍了与线程相关的概念，以及线程的概念模型，并在线程概念模型的基础上重点讲解了实现线程体的两种方式；然后介绍了 Java 线程的 4 种状态、线程的调度及线程优先级的作用，同时讲解两种特殊的线程：守护线程和主线程；最后利用实例说明线程同步的必要性和实现线程同步的两种方式。

5.10 练习题

1. 填空题

（1）每个进程都有 _____ 的代码和数据空间或称为进程上下文，进程切换的开销 _____ 。而线程可以看作是轻量级的进程，同一类线程 _____ 代码和数据空间，每个线程有 _____ 的运行栈和程序计数器，线程切换与进程切换相比开销要 _____ 。

（2）Java 中的线程由三部分组成：_____ 、_____ 和 _____ 。

（3）线程有 4 种状态：_____ 、_____ 、_____ 和 _____ 。

（4）Java 中线程有 _____ 个优先级由低到高分别用 _____ ~ _____ 表示。

（5）线程默认都是非守护线程，非守护线程也称作 _____ 。

2. 选择题

（1）Thread.NORM_PRIORITY 对应的级别数是 _____ 。

  A. 0   B. 1   C. 5   D. 10

（2）Thread thread = new Thread(); 如果要将 thread 设置为守护线程，那么应该如何编写代码。请选择 _____ 。

  A. thread.setDaemon(true)   B. thread.setDaemon(1)

  C. thread.setDaemon(False)   D. thread.setDaemon(0)

（3）下列哪种线程是 JVM 自动创建的 _____。

    A. 守护线程   B. 主线程   C. 非守护线程   D. 用户线程

（4）实现线程体的方式除了继承 Thread 类外，还可以实现以下哪个接口 _____。

    A. Cloneable   B. Runnable   C. Iterable   D. Serializable

（5）Java 中实现线程同步的关键字是 _____。

    A. static   B. final   C. synchronized   D. protected

3. 问答题

（1）简述线程和进程的关系。

（2）什么是守护线程？

（3）什么是主线程？主线程的特点是什么？

（4）简述实现线程体两种方式的使用范围。

5.11 编程练习

参照本章中的示例，分别用两种不同的方式实现线程体。

第 6 章
Java 集合框架

Java 集合框架（Java Collections Framework，JCF）提供了处理一组对象标准而高效的解决方案。严格地说，Java 集合框架出现在 Java 1.2 之后，包含设计精巧的数据结构和算法，便于开发者将主要精力放在业务功能实现上，从而减少底层设计的时间。

Java 集合框架在设计时大量使用了接口和抽象类，使得集合框架具有良好的扩展性。接口、接口的实现和集合算法是 Java 集合框架的三个主要组成部分。本章主要介绍 Java 集合框架中常用的接口和接口实现。

6.1 常用集合接口

Java 集合框架如图 6-1 所示。

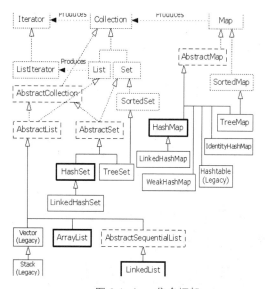

图 6-1 Java 集合框架

6.1.1 Collection 接口

Collection 接口是整个 Java 集合框架的基石，定义了集合框架中一些基本的方法。在某种意义上可以把 Collection 看成是动态的数组，一个对象的容器。通常把放入 Collection 中的对象称作元素。

Collection 接口的声明如下：

```
public interface Collection
```

Collection 接口的方法如下：

```
public boolean add(Object o)
```

说明：将对象添加进集合。

```
public boolean addAll(Collection c)
```

说明：将集合 c 中所有元素添加给此集合。

```
public void clear()
```

说明：删除集合中所有元素。

```
public boolean contains(Object o)
```

说明：查找集合中是否含有对象 o。

```
public boolean containsAll(Collection c)
```

说明：查找集合中是否含有集合 c 中所有元素。

```
public boolean equals(Object o)
```

说明：判断集合是否等价。

```
public int hashCode()
```

说明：返回集合的哈希码。

```
public boolean isEmpty()
```

说明：判断集合中是否有元素。

```
public Iterator iterator()
```

说明：返回一个迭代器，用于访问集合中的各个元素。

```
public boolean remove(Object o)
```

说明：如果集合中有与 o 相匹配的对象，就删除对象 o。

```
public boolean removeAll(Collection c)
```

说明：从集合中删除集合 c 中所有元素。

```
public boolean retainAll(Collection c)
```

说明：从集合中删除集合 c 中不包含的元素。

```
public int size()
```

说明：返回当前集合中元素的数量。

```
public Object[] toArray()
```

说明：以数组的形式返回集合中的元素。

```
public Object[] toArray(Object[] a)
```

说明：以数组的形式返回集合中与数组 a 类型匹配的元素。

6.1.2 List 接口

List 接口继承 Collection 接口并在其基础上进行扩充，有两个非常重要的实现类：ArrayList 和 LinkedList。

List 接口的声明如下：

```
public interface List extends Collection
```

List 接口的方法如下：

```
public void add(int index,Object element)
```

说明：在指定位置 index 上添加元素 element。

```
public boolean addAll(int index,Collection c)
```

说明：将集合 c 中的所有元素添加到指定位置 index。

```
public Object get(int index)
```

说明：返回列表中指定位置的元素。

```
public int indexOf(Object o)
```

说明：返回第一个出现元素 o 的位置，否则返回 -1。

```
public int lastIndexOf(Object o)
```

说明：返回最后一个出现元素 o 的位置，否则返回 -1。

```
public ListIterator listIterator()
```

说明：返回一个列表迭代器，用于访问列表中的元素。

```
public ListIterator listIterator(int index)
```

说明：返回一个列表迭代器，用于从指定位置 index 开始访问列表中的元素。

```
public Object remove(int index)
```

说明：删除指定位置上的元素。

```
public Object set(int index,Object element)
```

说明：用元素 element 取代位置 index 上的元素，并返回旧的元素。

```
public List subList(int fromIndex,int toIndex)
```

说明：返回从指定位置 fromIndex 到 toIndex 范围的子列表。

6.1.3 Set 接口

Set 接口继承 Collection 接口，自身并没有引入新的方法，只是不允许集合中存在相同的元素。每个具体的 Set 实现类依赖添加对象的 equals() 方法检查唯一性。

6.1.4 Map 接口

Map 接口不继承 Collection 接口。可以把 Map 看作是用于存储关键字/值对映射的容器，与 Set 接口相似，一个 Map 容器的关键字对象不允许重复。只有保证关键字对象的唯一性，才能完成根据关键字对象获取值对象的功能。

Map 接口的声明如下：

```
public interface Map
```

Map 接口的方法如下：

```
public void clear()
```

说明：从映像中删除所有映射。

```
public boolean containsKey(Object key)
```

说明：判断映像中是否存在关键字 key。

```
public boolean containsValue(Object value)
```

说明：判断映像中是否存在值 value。

```
public Set entrySet()
```

说明：返回 Map.Entry 对象的视图集，即映像中的关键字/值对。

```
public boolean equals(Object o)
```

说明：判断映像是否等价。

```
public Object get(Object key)
```

说明：获得与关键字 key 相关的值，并返回与关键字 key 相关的对象。如果没有在映像中找到该关键字，就返回 null。

```
public int hashCode()
```

说明：返回映像的哈希码。

```
public boolean isEmpty()
```

说明：判断映像中是否有映射。

```
public Set keySet()
```

说明：返回映像中所有关键字的视图集。

```
public Object put(Object key,Object value)
```

说明：将互相关联的一个关键字与一个值放入该映像。如果该关键字已经存在，那么与此关键字相关的新值将取代旧值。方法返回关键字的旧值，如果关键字原来并不存在，就返回 null。

```
public void putAll(Map t)
```

说明：将映像 t 的所有元素添加给该映像。

```
public Object remove(Object key)
```

说明：从映像中删除与 key 相关的映射。

```
public int size()
```

说明：返回当前映像中映射的数量。

```
public Collection values()
```

说明：返回映像中所有值的视图集。

6.1.5 Map.Entry 接口

如果说 Map 接口是一个用于存储关键字 / 值对映射的容器，那么 Map.Entry 接口就是用于描述存储在 Map 容器中的一个关键字 / 值对映射。

Map.Entry 接口的声明如下：

```
public static interface Map.Entry
```

Map.Entry 接口的方法如下：

```
public Object getKey()
```

说明：返回映像的关键字。

```
public Object getValue()
```

说明：返回映像的值。

```
public Object setValue(Object value)
```

说明：设置映像的值。

6.1.6 Iterator 接口

Iterator 接口用于对集合容器进行向前的单方向遍历，通常称作迭代器。

Iterator 接口的声明如下：

```
public interface Iterator
```

Iterator 接口的方法如下：

```
public boolean hasNext()
```

说明：判断是否存在下一个元素。

```
public Object next()
```

说明：返回下一个元素。如果到达集合结尾，就抛出 NoSuchElementException 异常。

```
public void remove()
```

说明：删除刚访问的元素。

6.1.7 ListIterator 接口

ListIterator 接口与 Iterator 接口类似，不同之处在于 ListIterator 用于对 List 容器进行双向遍历。

ListIterator 接口的声明如下：

```
public interface ListIterator extends Iterator
```

ListIterator 接口的方法如下：

```
public void add(Object o)
```

说明：将对象 o 添加到当前位置。

```
public boolean hasNext()
```

说明：判断是否存在下一个元素。

```
public boolean hasPrevious()
```

说明：判断是否存在上一个元素。

```
public Object next()
```

说明：返回下一个元素。

```
public int nextIndex()
```

说明：返回下一个索引位置。

```
public Object previous()
```

说明：返回上一个元素。

```
public int previousIndex()
```

说明：返回上一个索引位置。

```
public void remove()
```

说明：删除刚访问的元素。

```
public void set(Object o)
```

说明：用对象 o 取代刚访问的元素。

 在进行程序设计时，建议使用接口作为方法的返回值。

6.2 常用集合类

本小节介绍各接口的实现类，包括 ArrayList、LinkedList、HashSet 类的声明、构造方法及使用。

6.2.1 ArrayList 类

ArrayList 类的声明如下：

```
public class ArrayList extends AbstractList implements List, RandomAccess, Cloneable, Serializable
```

ArrayList 类是 List 接口的重要实现类，就像一个动态分配的数组，每次将新添加的对象放在最后。

ArrayList 类的构造方法如下：

```
public ArrayList()
public ArrayList(Collection c)
public ArrayList(int initialCapacity)
```

ArrayList 类的主要方法如下：

```
public void add(int index,Object element)
```

说明：在指定位置 index 上添加元素 element。

```
public boolean add(Object o)
```

说明：将对象添加到 ArrayList 的最后。

```
public boolean addAll(Collection c)
```

说明：将集合 c 中所有元素添加到 ArrayList 的最后。

```
public boolean addAll(int index,Collection c)
```

说明：将集合 c 中的所有元素添加到 ArrayList 指定位置 index 的后面。

```
public void clear()
```

说明：删除 ArrayList 中所有元素。

```
public Object clone()
```

说明：克隆 ArrayList 对象。

```
public boolean contains(Object elem)
```

说明：查找 ArrayList 中是否含有对象 elem。

```
public void ensureCapacity(int minCapacity)
```

说明：将 ArrayList 对象容量增加 minCapacity。

`public Object get(int index)`

说明：返回 ArrayList 中指定位置的元素。

`public int indexOf(Object elem)`

说明：返回第一个出现元素 elem 的位置，否则返回 -1。

`public boolean isEmpty()`

说明：判断 ArrayList 中是否有元素。

`public Iterator iterator()`

说明：返回一个迭代器，用于访问 ArrayList 中的各个元素。

`public int lastIndexOf(Object elem)`

说明：返回最后一个出现元素 elem 的位置，否则返回 -1。

`public Object remove(int index)`

说明：删除指定位置上的元素。

`public Object set(int index,Object element)`

说明：利用元素 element 取代位置 index 上的元素，并返回旧的元素。

`public int size()`

说明：返回当前 ArrayList 中元素的数量。

`public Object[] toArray()`

说明：以数组的形式返回 ArrayList 中的元素。

`public Object[] toArray(Object[] a)`

说明：以数组的形式返回 ArrayList 中与数组 a 类型匹配的元素。

`public void trimToSize()`

说明：设置 ArrayList 对象容量为列表当前大小。

应用示例如下：

```
import java.util.ArrayList;
import java.util.Iterator;
public class ArrayListDemo {
    public static void main(String[] args) {
```

```java
        ArrayList list1 = new ArrayList();
        list1.add("One");
        list1.add("Two");
        list1.add("Three");
        list1.add(0, "Zero");
        System.out.println("<----list1 中共有 " + list1.size() +
" 个元素 >");
        System.out.println("<--list1 中的内容: " + list1 + "-->");
        ArrayList list2 = new ArrayList();
        list2.add("Begin");
        list2.addAll(list1);
        list2.add("End");
        System.out.println("<----list2 中共有 " + list2.size() +
" 个元素 >");
        System.out.println("<--list2 中的内容: " + list2 + "-->");
        ArrayList list3 = new ArrayList(list2);
        list3.removeAll(list1);
        System.out.println("<--list3 中是否存在 One: "
                + (list3.contains("one") ? "是" : "否") + "-->");
        list3.add(1, "same element");
        list3.add(2, "same element");
        System.out.println("<----list3 中共有 " + list3.size() +
" 个元素 >");
        System.out.println("<--list3 中的内容: " + list3 + "-->");
        System.out.println("<--list3 中第一次出现 same element 的索引是 "
                + list3.indexOf("same element") + "-->");
        System.out.println("<--list3 中最后一次出现 same element 的索引是 "
                + list3.lastIndexOf("same element") + "-->");
        System.out.println("<-- 使用 Iterator 接口访问 list3-->");
        Iterator it = list3.iterator();
        while (it.hasNext()) {
            String str = (String) it.next();
            System.out.println("<--list3 中的元素:"+str+"-->");
        }
        System.out.println("<-- 将 list3 中的 same element 修改为 another element-->");
        list3.set(1, "another element");
        list3.set(2, "another element");
        System.out.println("<-- 将 list3 转为数组 -->");
        Object[] array = list3.toArray();
        for (int i = 0; i < array.length; i++) {
            String str = (String) array[i];
            System.out.println("array[" + i + "]=" + str);
        }
```

```
            System.out.println("<--清空list3-->");
            list3.clear();
            System.out.println("<--list3 中是否为空: " + (list3.isEmpty()
? "是" : "否") + "-->");
            System.out.println("<----list3 中共有" + list3.size() +
"个元素>");
    }
}
```

这段程序代码的运行结果如下:

```
<----list1 中共有 4 个元素 >
<--list1 中的内容: [Zero, One, Two, Three]-->
<----list2 中共有 6 个元素 >
<--list2 中的内容: [Begin, Zero, One, Two, Three, End]-->
<--list3 中是否存在 One: 否 -->
<----list3 中共有 4 个元素 >
<--list3 中的内容: [Begin, same element, same element, End]-->
<--list3 中第一次出现 same element 的索引是 1-->
<--list3 中最后一次出现 same element 的索引是 2-->
<-- 使用 Iterator 接口访问 list3-->
<--list3 中的元素: Begin-->
<--list3 中的元素: same element-->
<--list3 中的元素: same element-->
<--list3 中的元素: End-->
<-- 将 list3 中的 same element 修改为 another element-->
<-- 将 list3 转为数组 -->
array[0]=Begin
array[1]=another element
array[2]=another element
array[3]=End
<-- 清空 list3-->
<--list3 中是否为空: 是 -->
<----list3 中共有 0 个元素 >
```

6.2.2 LinkedList 类

LinkedList 类的声明如下:

```
public class LinkedList extends AbstractSequentialList implements List,
Cloneable, Serializable
```

LinkedList 类是 List 接口的另一个重要的实现类。因为 LinkedList 类的底层实现是链表,所以便于将新加入的对象插入指定位置。

LinkedList 类的构造方法如下：

```
public LinkedList()
public LinkedList(Collection c)
```

LinkedList 类的主要方法如下：

```
public void add(int index,Object element)
```

说明：在指定位置 index 上添加元素 element。

```
public boolean add(Object o)
```

说明：将对象添加到 LinkedList 的最后。

```
public boolean addAll(Collection c)
```

说明：将集合 c 中所有元素添加到 LinkedList 的最后。

```
public boolean addAll(int index,Collection c)
```

说明：将集合 c 中所有元素添加到 LinkedList 指定位置 index 的后面。

```
public void addFirst(Object o)
```

说明：将对象 o 添加到 LinkedList 的第一位。

```
public void addLast(Object o)
```

说明：将对象 o 添加到 LinkedList 的最后一位。

```
public void clear()
```

说明：删除 LinkedList 中所有元素。

```
public Object clone()
```

说明：克隆 LinkedList 对象。

```
public boolean contains(Object o)
```

说明：查找 LinkedList 中是否含有对象 o。

```
public Object get(int index)
```

说明：返回 LinkedList 中指定位置的元素。

```
public Object getFirst()
```

说明：返回列表第一位的元素。

```
public Object getLast()
```

说明：返回列表最后一位的元素。

```
public int indexOf(Object o)
```

说明：返回第一个出现元素 o 的位置，否则返回 -1。

```
public ListIterator listIterator(int index)
```

说明：返回一个列表迭代器，用于从指定位置 index 开始访问列表中的元素。

```
public int lastIndexOf(Object o)
```

说明：返回最后一个出现元素 o 的位置，否则返回 -1。

```
public Object remove(int index)
```

说明：删除指定位置上的元素。

```
public boolean remove(Object o)
```

说明：如果 LinkedList 中有与 o 相匹配的对象，就删除对象 o。

```
public Object removeFirst()
```

说明：删除并返回 LinkedList 中第一个元素。

```
public Object removeLast()
```

说明：删除并返回 LinkedList 中最后一个元素。

```
public Object set(int index,Object element)
```

说明：利用元素 element 取代位置 index 上的元素，并返回旧的元素。

```
public int size()
```

说明：返回当前 LinkedList 中元素的数量。

```
public Object[] toArray()
```

说明：以数组的形式返回 LinkedList 中的元素。

```
public Object[] toArray(Object[] a)
```

说明：以数组的形式返回 LinkedList 中与数组 a 类型匹配的元素。

应用示例如下：

```
import java.util.LinkedList;
```

```java
import java.util.ListIterator;
public class LinkedListDemo {
    public static void main(String[] args) {
        LinkedList list = new LinkedList();
        list.add("One");
        list.add("Two");
        list.add("Three");
        System.out.println("<--list 中共有 " + list.size() +
 " 个元素 -->");
        System.out.println("<--list 中的内容: " + list + "-->");
        String first = (String) list.getFirst();
        String last = (String) list.getLast();
        System.out.println("<--list 的第一个元素是 " + first +
 "-->");
        System.out.println("<--list 的最后一个元素是 " + last +
 "-->");
        list.addFirst("Begin");
        list.addLast("End");
        System.out.println("<--list 中共有 " + list.size() +
 " 个元素 -->");
        System.out.println("<--list 中的内容: " + list + "-->");
        System.out.println("<-- 使用 ListIterator 接口操作 ist-->");
        ListIterator lit = list.listIterator();
        System.out.println("<-- 下一个索引是 " + lit.nextIndex() +
            "-->");
        lit.next();
        lit.add("Zero");
        lit.previous();
        lit.previous();
        System.out.println("<-- 上一个索引是 " + lit.previousIndex()
 + "-->");
        lit.set("Start");
        System.out.println("<--list 中的内容: " + list + "-->");
        System.out.println("<-- 删除 list 中的 Zero-->");
        lit.next();
        lit.next();
        lit.remove();
        System.out.println("<--list 中的内容: " + list + "-->");
        System.out.println("<-- 删除 list 中第一个和最后一个元素
-->");
        list.removeFirst();
        list.removeLast();
        System.out.println("<--list 中共有 " + list.size() +
            " 个元素 -->");
```

```
            System.out.println("<--list 中的内容: " + list + "-->");
    }
}
```

这段程序代码的运行结果如下:

```
<--list 中共有 3 个元素 -->
<--list 中的内容: [One, Two, Three]-->
<--list 的第一个元素是 One-->
<--list 的最后一个元素是 Three-->
<--list 中共有 5 个元素 -->
<--list 中的内容: [Begin, One, Two, Three, End]-->
<-- 使用 ListIterator 接口操作 list-->
<-- 下一个索引是 0-->
<-- 上一个索引是 -1-->
<--list 中的内容: [Start, Zero, One, Two, Three, End]-->
<-- 删除 list 中的 Zero-->
<--list 中的内容: [Start, One, Two, Three, End]-->
<-- 删除 list 中第一个和最后一个元素 -->
<--list 中共有 3 个元素 -->
<--list 中的内容: [One, Two, Three]-->
```

6.2.3 HashSet 类

HashSet 类的声明如下:

```
public class HashSet extends AbstractSet implements Set, Cloneable, Serializable
```

HashSet 类是 Set 接口实现类中最常用的一个。因为 HashSet 类通过 Hash 算法进行存储,所以具有快速定位元素的特点。

HashSet 类的构造方法如下:

```
public HashSet()
public HashSet(Collection c)
public HashSet(int initialCapacity,float loadFactor)
public HashSet(int initialCapacity)
```

HashSet 类的主要方法如下:

```
public boolean add(Object o)
```

说明:将对象添加进 HashSet。

```
public void clear()
```

说明:删除 HashSet 中所有元素。

```
public Object clone()
```

说明：克隆 HashSet 对象。

```
public boolean contains(Object o)
```

说明：查找 HashSet 中是否含有对象 o。

```
public boolean isEmpty()
```

说明：判断 HashSet 中是否有元素。

```
public Iterator iterator()
```

说明：返回一个迭代器，用于访问 HashSet 中的各个元素。

```
public boolean remove(Object o)
```

说明：如果 HashSet 中有与 o 相匹配的对象，就删除对象 o。

```
public int size()
```

说明：返回当前 HashSet 中元素的数量。

应用示例如下：

```java
import java.util.HashSet;
public class HashSetDemo {
    public static void main(String[] args) {
        HashSet set1 = new HashSet();
        set1.add("One");
        set1.add("Two");
        set1.add("Three");
        set1.add("Zero");
        set1.add("One");
        System.out.println("<--set1 中的内容: " + set1 + "-->");
        HashSet set2 = new HashSet();
        set2.add("Zero");
        set2.add("Four");
        System.out.println("<--set2 中的内容: " + set2 + "-->");
        System.out.println("<-- 从 set1 中删除 set2 中包含的元素 -->");
        set1.removeAll(set2);
        System.out.println("<--set1 中的内容: " + set1 + "-->");
        System.out.println("<--set1 中是否存在 One: "
                + (set1.contains("One") ? "是" : "否") + "-->");
        System.out.println("<-- 清空 set1-->");
        set1.clear();
        System.out.println("<--set1 中是否为空: " + (set1.isEmpty() ?
```

```
            "是" : "否") + "-->");
            System.out.println("<----set1 中共有 " + set1.size() +
"个元素>");
    }
}
```

这段程序代码的运行结果如下：

```
<--set1 中的内容：[Three, Two, Zero, One]-->
<--set2 中的内容：[Four, Zero]-->
<-- 从 set1 中删除 set2 中包含的元素 -->
<--set1 中的内容：[Three, Two, One]-->
<--set1 中是否存在 One：是 -->
<-- 清空 set1-->
<--set1 中是否为空：是 -->
<----set1 中共有 0 个元素 >
```

6.2.4 HashMap

HashMap 类的声明如下：

```
public class HashMap extends AbstractMap implements Map, Cloneable,
Serializable
```

HashMap 是 Map 接口的重要实现类，在 Java 程序设计中经常用到。因为 HashMap 也采用 Hash 算法，所以可以快速定位关键字对象。

HashMap 类的构造方法如下：

```
public HashMap()
public HashMap(Map m)
public HashMap(int initialCapacity,float loadFactor)
public HashMap(int initialCapacity)
```

HashMap 类的主要方法如下：

```
public void clear()
```

说明：删除 HashMap 中所有元素。

```
public Object clone()
```

说明：克隆 HashMap 对象。

```
public boolean containsKey(Object key)
```

说明：判断映像中是否存在关键字 key。

```
public boolean containsValue(Object value)
```

说明:判断映像中是否存在值 value。

```
public Set entrySet()
```

说明:返回 Map.Entry 对象的视图集,即映像中的关键字/值对。

```
public Object get(Object key)
```

说明:获得与关键字 key 相关的值,并返回与关键字 key 相关的对象。如果没有在映像中找到该关键字,就返回 null。

```
public boolean isEmpty()
```

说明:判断映像中是否有映射。

```
public Set keySet()
```

说明:返回映像中所有关键字的视图集。

```
public Object put(Object key,Object value)
```

说明:将互相关联的一个关键字与一个值放入该映像。如果该关键字已经存在,那么与此关键字相关的新值将取代旧值。方法返回关键字的旧值,如果关键字原来并不存在,就返回 null。

```
public void putAll(Map t)
```

说明:将映像 t 的所有元素添加给该映像。

```
public Object remove(Object key)
```

说明:从映像中删除与 key 相关的映射。

```
public int size()
```

说明:返回当前映像中映射的数量。

```
public Collection values()
```

说明:返回映像中所有值的视图集。

应用示例如下:

```
import java.util.Collection;
import java.util.HashMap;
import java.util.Iterator;
import java.util.Set;
```

```java
import java.util.Map;
public class HashMapDemo {
    public static void main(String[] args) {
        HashMap hm = new HashMap();
        hm.put("1", "January");
        hm.put("2", "February");
        hm.put("3", "March");
        hm.put("4", "April");
        hm.put("5", "May");
        hm.put("6", "June");
        hm.put("7", "July");
        hm.put("8", "August");
        hm.put("9", "September");
        hm.put("10", "October");
        hm.put("11", "November");
        hm.put("12", "December");
        System.out.println("<--hm 中是否存在值为 November 的映射: "       +
(hm.containsValue("November") ? "是" : "否") + "-->");
        System.out.println("<--hm 中是否存在关键字为 13 的映射: "
                    + (hm.containsKey("13") ? "是" : "否") + "-->");
        System.out.println("<--hm 中的关键字有: -->");
        Set keys = hm.keySet();
        Iterator kit = keys.iterator();
        while (kit.hasNext()) {
            String key = (String) kit.next();
            System.out.println("<-- 关键字: " + key + "-->");
        }
        System.out.println("<--hm 中的值有: -->");
        Collection values = hm.values();
        Iterator vit = values.iterator();
        while (vit.hasNext()) {
            String value = (String) vit.next();
            System.out.println("<-- 值: " + value + "-->");
        }
        System.out.println("<--hm 中的映射为: -->");
        Set set = hm.entrySet();
        Iterator it = set.iterator();
        while (it.hasNext()) {
            Map.Entry me = (Map.Entry) it.next();
            System.out.println("[" + me.getKey() + "," + me.getValue()
 + "]");
        }
    }
}
```

这段程序代码的运行结果如下：

```
<--hm 中是否存在值为 November 的映射：是 -->
<--hm 中是否存在关键字为 13 的映射：否 -->
<--hm 中的关键字有：-->
<-- 关键字：10-->
<-- 关键字：3-->
<-- 关键字：5-->
<-- 关键字：7-->
<-- 关键字：2-->
<-- 关键字：11-->
<-- 关键字：9-->
<-- 关键字：4-->
<-- 关键字：8-->
<-- 关键字：12-->
<-- 关键字：6-->
<-- 关键字：1-->
<--hm 中的值有：-->
<-- 值：October-->
<-- 值：March-->
<-- 值：May-->
<-- 值：July-->
<-- 值：February-->
<-- 值：November-->
<-- 值：September-->
<-- 值：April-->
<-- 值：August-->
<-- 值：December-->
<-- 值：June-->
<-- 值：January-->
<--hm 中的映射为：-->
[10,October]
[3,March]
[5,May]
[7,July]
[2,February]
[11,November]
[9,September]
[4,April]
[8,August]
[12,December]
[6,June]
[1,January]
```

 在选择使用集合类时，建议开发者深入了解其自身的特点和适用范围。

6.3 实例练习：集合类的综合运用

在本章的最后，我们将综合运用前面所学的知识给出一个完整的实例。

该实例综合运用 Collection、Set、Map、Map.Entry 和 Iterator 接口，以及 ArrayList 和 HashMap 类，实现简单的取模分组并打印显示的功能。实例的源代码如下：

```java
import java.util.ArrayList;
import java.util.Collection;
import java.util.HashMap;
import java.util.Iterator;
import java.util.Map;
import java.util.Set;
public class Integration {
    /**
     * 取模分组
     * @param data
     * @param mod
     * @return Map
     */
    public static Map grouping(int[] data, int mod) {
        Map map = new HashMap();
        for (int i = 0; i < data.length; i++) {
            Integer key = new Integer(data[i] % mod);
            if (map.containsKey(key)) {
                Collection col = (ArrayList) map.get(key);
                col.add(new Integer(data[i]));
            } else {
                Collection col = new ArrayList();
                col.add(new Integer(data[i]));
                map.put(key, col);
            }
        }
        return map;
    }
    /**
     * 打印显示
     * @param map
     */
    public static void printMap(Map map) {
        Set set = map.entrySet();
        Iterator outerIt = set.iterator();
        while (outerIt.hasNext()) {
            Map.Entry me = (Map.Entry) outerIt.next();
```

```
                Integer key = (Integer) me.getKey();
                System.out.println("<--第" + key + "组成员列表开始-->");
                Collection col = (Collection) me.getValue();
                Iterator innerIt = col.iterator();
                while (innerIt.hasNext()) {
                    System.out.println("成员编号: " + innerIt.next());
                }
                System.out.println("<--第" + key + "组成员列表结束-->");
            }
        }
        public static void main(String[] args) {
            int[] data = { 13, 24, 59, 78, 30, 14, 32, 6, 5, 81, 48 };
            Map map = Integration.grouping(data, 3);
            Integration.printMap(map);
        }
    }
```

6.4 要点总结

本章首先概述了 Java 集合框架的内容和框架继承结构；然后重点介绍了 Java 集合框架中重要和常用的接口；最后以示例的方式介绍了具体集合类的使用方法和特点。

6.5 练习题

1. 填空题

（1）ArrayList 类通过 _____ 方法获得迭代器，从而对所有元素进行遍历。

（2）Map 接口通过 _____ 方法返回 Map.Entry 对象的视图集，即映像中的关键字/值对。

（3）ArrayList 类通过 _____ 方法增加容量。

（4）ArrayList 类通过 _____ 方法设置其容量为列表当前大小。

（5）HashMap 类通过 _____ 方法返回映像中所有值的视图集。

2. 选择题

（1）下列集合类中，哪个可以使用 ListIterator 进行遍历。_____

    A. ArrayList   B. List   C. HashMap   D. LinkedList

（2）下列哪个集合类使用链表作为底层实现的方式。_____

    A. ArrayList   B. LinkedList   C. HashSet   D. HashMap

（3）下列哪个集合类可以用于存储关键字/值对映像。_____

    A. Map   B. Map.Entry   C. HashMap   D. HashSet

（4）下列哪个集合类不允许存在相同的元素。_____

    A. HashSet   B. Set   C. ArrayList   D. LinkedList

（5）下列哪个接口没有继承 Collection 接口。_____

    A. Map   B. HashMap   C. Set   D. List

3. 问答题

（1）简述 Java 集合的作用。

（2）简述 Collection、Map、List、HashMap、Set、ArrayList 的关系。

6.6 编程练习

结合本章所学内容编写一个方法，实现将存储在 HashMap 中的数据转存到 Collection 类型的聚集中。

第 7 章
Java IO

Java 的核心库 java.io 提供了全面的 IO 接口，包括文件读写、标准设备输出等。Java 中 IO 是以流为基础进行输入 / 输出的，所有数据被串行化写入输出流，或者从输入流读入。本章主要介绍 Java 语言的输入 / 输出技术。

7.1 File 类

File 类是一个与流无关的类，可以进行创建或删除文件等常用操作。使用 File 类，首先要掌握其构造方法。常用的构造方法如下：

```
public File(String pathname)        pathname --- 路径名字符串
```

实例化 File 类时，必须设置好路径，即向 File 类的构造方法中传递一个文件路径。例如，要操作 D 盘中的 IODemo.java 文件，必须把 pathname 写成"d:\\IODemo.java"，其中"\\"表示目录的分隔符。Java 的 File 类定义了两个常量使程序可以在任意操作系统中使用。

（1）pathSeqarator：与系统有关的路径分隔符字符串；

（2）separator：与系统有关的默认名称分隔符字符串。

File 类操作文件的常用方法如下：

```
import java.io.File;
```

```java
public class FileDemo01 {
    public static void main(String[] args) {
        System.out.println("路径分隔符: " + File.separator);
        File file = new File("D:\\sn.txt");
        // 建议使用如下路径分隔符, 必须给出完整的路径
        // File file = new File("D:" + File.separator + "sn.txt");
        if (file.exists()) {                    // 判断创建的文件是否存在
            file.delete();                      // 删除文件
        } else {
            try {
                file.createNewFile();  // 创建文件
            } catch (Exception e) {
                System.out.println("创建文件失败! ");
            }
        }
    }
}
```

上面的示例演示了创建和删除文件的方法,建议在操作文件时一定要使用 File.separator 表示分隔符,使开发的程序在任何操作系统中都能使用。

下面的示例演示了 File 类对文件夹的操作,创建路径与文件操作实例化 File 类一样,是利用 mkdir() 方法完成创建文件夹的。

```java
import java.io.File;
public class FileDemo02 {
    public static void main(String[] args) {
        // 给出文件夹的完整路径
        File fdir = new File("d:" + File.separator + "jkx");
        fdir.mkdir();                           // 创建文件夹
        // 列出指定目录的全部内容 (目录和文件)
        String[] str = fdir.list();        // 返回一个字符串数组
        for (int i = 0; i < str.length; i++) {
            System.out.println(str[i]); // 列出给定目录中的内容
        }
        System.out.println("常用的方法是: listFiles() 列出完整的路径");
        File[] files = fdir.listFiles();   // 列出全部文件
        for (int i = 0; i < files.length; i++) {
            System.out.println(files[i]);
        }
        if (fdir.isDirectory()) {               // 判断是否是目录
            System.out.println(fdir.getPath() + " 路径是目录");
        } else {
            System.out.println(fdir.getPath() + " 路径不是目录");
```

```
        }
    }
}
```

程序运行结果如下：

```
rjjs
javaIO 实验 .doc
09 软件工程
相关知识点 .txt
常用的方法是：listFiles() 列出完整的路径
d:\jkx\rjjs
d:\jkx\javaIO 实验 .doc
d:\jkx\09 软件工程
d:\jkx\ 相关知识点 .txt
d:\jkx 路径是目录
```

7.2 RandomAccessFile 类

File 类只对文件本身进行操作，而如果要对文件内容进行操作，就应当使用 RandomAccessFile 类。此类属于随机读取类，即可以随机地读取一个文件中指定位置的数据。既不是输入流的子类，也不是输出流的子类。

RandomAccessFile 类常用的构造方法有以下两种。

1. public RandomAccessFile(String file,String mode)

name：和系统相关的文件名。

mode：对文件的访问权限，可以是 r、rw、rws 或 rwd。

2. public RandomAccessFile(File file,String mode)

file：一个 File 类的对象。

mode：对文件的访问权限，可以是 r、rw、rws 或 rwd。

利用构造方法显示文件本身源代码的执行过程如下：

```
import java.io.File;
import java.io.RandomAccessFile;
public class RandomAccessFileDemo01 {
    public static void main(String[] args) throws Exception{
        File f = new File("d:" + File.separator + "kj.txt");
        // 创建随机访问文件为读写
        RandomAccessFile raf = new RandomAccessFile(f,"rw");
        long filePoint = 0;                           // 定义循环变量
```

```
            long fileLength = raf.length();           // 获取文件长度
            while (filePoint<fileLength) {
                String str = raf.readLine();          // 从文件中按行读取
                System.out.println(str);
                filePoint = raf.getFilePointer();
            }
            raf.close();                              // 关闭文件
    }
}
```

下面的示例演示了 D 盘存在文件 "kj.txt"，创建 int 型数组，把 int 型数组写入到文件 "kj.txt" 中，然后按倒序读出这些数据。

```
import java.io.File;
import java.io.RandomAccessFile;
public class RandomAccessFileDemo02 {
    public static void main(String[] args) throws Exception{
        int[] score = {67,60,90,70,53,78};
        // 创建随机访问文件为读写
        RandomAccessFile raf = new
RandomAccessFile("d:"+File.separator+"kj.txt","rw");
        for (int i = 0; i < score.length; i++) {
            raf.writeInt(score[i]);
        }
        for (int i = score.length-1; i >= 0; i--) {
            raf.seek(i*4);                            // 整型占 4 个字节
            System.out.print(raf.readInt()+"\t");
        }
        raf.close();                                  // 关闭文件
    }
}
```

程序运行结果如下：

78 53   70   90   60   67

虽然随机读写流可以实现对文件内容的操作，但是却过于复杂。一般情况下，操作文件内容往往会使用字节流或字符流。

7.3 字节流与字符流

流（stream）是一组有序的数据序列。根据操作的类型，分为输入流和输出流两种。在程序中，所有的数据流都是以流的方式进行传输和保存的。程序需要数据时，就使用输入流读取数据,程序需要将一些数据保存起来时,就使用输出流。

流的操作主要有字节流和字符流两大类,两类都有输入和输出操作。在字节流中输入数据使用的是 InputStream 类,输出数据使用的是 OutputStream 类。在字符流中输入数据使用是 Reader 类,输出数据使用是 Writer 类。

7.3.1 字节流

字节流主要操作 byte 类型数据。以 byte 数组为准,主要操作类是 InputStream 和 OutputStream。

1. 字节输出流:OutputStream 类

OutputStream 类是一个抽象类,如果使用此类,就必须通过子类实例化对象。如果要操作的对象是一个文件,就可以使用 FileOutputStream 类,通过向上转型实现实例化。下面的示例演示了向文件中写入字符串。

```java
import java.io.File;
import java.io.FileOutputStream;
import java.io.OutputStream;
public class OutputStreamDemo {
  public static void main(String[] args) throws Exception{
      // 找到一个文件
      File f = new File("d:" + File.separator + "kj.txt");
      // 实例化 -- 子类实例化父类
      OutputStream out = new FileOutputStream(f);
      String str = "zknu.jkx.czw";
      byte[] b = str.getBytes();
      // 1. 循环把每一个字节一个个写入到文件中
      for (int i = 0; i < b.length; i++) {
          out.write(b[i]);
      }
      // 2. 将 byte 数组写入到文件中
      out.write(b);          // 内容保存
      out.close();           // 关闭输出流
  }
}
```

程序用两种方式将所有字节写入到文件中,但是如果重新运行程序,就会覆盖文件中已有的内容。如果是在原文件中追加内容,就使用下面的构造方法。

```
public FileOutputStream(File file,boolean append)
```

append 为 true,表示在文件的末尾追加内容。请读者自行测试。

2. 字节输入流:InputStream 类

InputStream 类可以把内容从文件中读取出来。InputStream 类本身是一个抽象类,必须依靠其子类实例化对象，对文件操作的子类为 FileInputStream。

下面的示例演示了对文件的读取方式。

```java
import java.io.File;
import java.io.FileInputStream;
import java.io.InputStream;
public class InputStreamDemo {
    public static void main(String[] args) throws Exception{
        // 找到一个文件
        File f = new File("d:" + File.separator + "kj.txt");
        // 实例化 -- 子类实例化父类
        InputStream in = new FileInputStream(f);
        byte[] b = new byte[1024];
        int len = in.read(b);
        // 1. 循环把每一个字节一个个写入到文件中
        for (int i = 0; i < len; i++) {
            b[i] = (byte)in.read();
        }
        // 2. 将 byte 数组写入到文件中
        in.close();              // 关闭输出流
        System.out.println(new String(b,0,len));
    }
}
```

7.3.2 字符流

Java 中的一个字符占两个字节。Java 提供了两个专门操作字符流的类：Reader 和 Writer。这两个类是字符流的抽象类，定义了字符流读取和写入的基本方法，各个子类会依其特点实现或覆盖这些方法。

1. 字符输出流：Writer 类

Writer 类对文件操作的子类是 FileWriter 类。

下面的示例演示了 Writer 类向文件中写入数据的方法。

```java
import java.io.File;
import java.io.FileWriter;
import java.io.Writer;
public class WriterDemo {
    public static void main(String[] args) throws Exception{
        // 找到一个文件
        File f = new File("d:" + File.separator + "kj.txt");
        // 实例化 -- 子类实例化父类
        Writer out = new FileWriter(f,true);
```

```
            String str = "\r\ncomputer engineering dept";
            out.write(str);     // 内容保存
            out.close();        // 关闭输出流
    }
}
```

2. 字符输入流：Reader 类

Reader 类对文件操作的子类是 FileReader 类。

下面的示例演示了 Reader 类从文件中读取数据的方法。

```
import java.io.File;
import java.io.FileReader;
import java.io.Reader;
public class ReaderDemo {
    public static void main(String[] args) throws Exception{
            // 找到一个文件
            File f = new File("d:" + File.separator + "kj.txt");
            // 实例化 -- 子类实例化父类
            Reader reader = new FileReader(f);
            int len = 0 ;// 用来记录读取的数据个数
            char[] c = new char[1024];  // 所有的内容读到此数组中
            int temp = 0; // 接收读取的每一个内容
            while ((temp = reader.read())!= -1) {
                    // 将每次读取的内容给 temp，temp 的值等于 -1，文件读完
                    c[len] = (char)temp;
                    len++;
            }
            reader.close();
            System.out.println(" 内容为： " + new String(c,0,len));
    }
}
```

7.3.3 字节流与字符流的区别

字节流与字符流在操作上非常类似，二者除了代码不同之外，最大的差别是操作时是否使用缓冲区（可理解为一段特殊的内存）。字节流在操作时不使用缓冲区，是文件本身直接操作的，而字符流在操作时使用缓冲区，即先通过缓冲区再操作文件。修改 WriterDemo.java 文件如下：

```
import java.io.File;
import java.io.FileWriter;
import java.io.Writer;
public class WriterDemo01 {
    public static void main(String[] args) throws Exception{
```

```
            // 找到一个文件
            File f = new File("d:" + File.separator + "kj.txt");
            // 实例化 -- 子类实例化父类
            Writer out = new FileWriter(f,true);
            String str = "\r\ncomputer engineering dept";
            out.write(str);                // 内容保存
            out.flush();                   // 刷新缓冲区
            // out.close();                // 关闭输出流
    }
}
```

如果不关闭输出流,就需要刷新缓冲区才能在文件中看到内容,读者可自行演示。因为所有文件在硬盘或在传输时都是以字节的方式进行的,而字符只有在内存中才会形成,所以在开发中字节流使用比较广泛。

7.4 转换流

Java JDK 文档中的 FileWriter 类并不直接是 Writer 类的子类,而是 OutputStreamWriter 的子类;FileReader 类并不直接是 Reader 类的子类,而是 InputStreamReader 的子类,两个子类中间都需要进行转换操作,这两个类就是字节流-字符流的转换类。

- OutputStreamWriter 类:将输出的字符流转换为字节流,即将一个字符流的输出对象变为字节流的输出对象。
- InputStreamReader 类:将输入的字节流转换为字符流,即将一个字节流的输入对象变为字符流的输入对象。

无论如何转换和操作,最终都是以字节的形式保存在文件中。例如:

将字节输出流变为字符输出流。

```
import java.io.File;
import java.io.FileOutputStream;
import java.io.OutputStreamWriter;
import java.io.Writer;
public class OutputStreamWriterDemo {
    public static void main(String[] args) throws Exception{
            File f = new File("d:" + File.separator + "kj.txt");
            // 实例化 -- 字节流变为字符流
            Writer out=new OutputStreamWriter(new FileOutputStream(f));
            out.write("www.zknu.edu.cn");
            out.close();
    }
```

}

将字节输入流变为字符输入流。

```java
import java.io.File;
import java.io.FileInputStream;
import java.io.InputStreamReader;
import java.io.Reader;
public class InputStreamReaderDemo {
    public static void main(String[] args) throws Exception{
        File f = new File("d:" + File.separator + "kj.txt");
        // 实例化 -- 字节流变为字符流
        Reader rd=new InputStreamReader(new FileInputStream(f));
        char[] c = new char[1024];
        int len = rd.read(c);
        rd.close();
        System.out.println("内容为: " + new String(c,0,len));
    }
}
```

7.5 打印流

在 Java io 包中，打印流提供了非常方便的打印功能，可以打印任何数据类型，如小数、整数、字符串等。打印流是输出信息最方便的类，主要包含字节打印流（PrintStream）和字符打印流（PrintWriter）。本节主要通过字节打印流（PrintStream）进行讲解。

字节打印流（PrintStream）是 OutputStream 类的子类，其中一个构造方法可以直接接收 OutputStream 类的实例，以便输出数据。例如：

```java
import java.io.File;
import java.io.FileOutputStream;
import java.io.PrintStream;
public class PrintStreamDemo {
    public static void main(String[] args) throws Exception{
        File f = new File("d:" + File.separator + "kj.txt");
        // 通过 FileOutputStream 实例化，向文件中打印输出
        PrintStream ps = new PrintStream(new FileOutputStream(f));
        ps.println("welcome you!");
        ps.println(3*3);
        ps.close();
    }
}
```

运行结果如图 7.2 所示。

图 7.2 运行结果

7.6 管道流

管道流的主要作用是可以进行两个线程间的通信，分为管道输出流（PipeOutputStream）和管道输入流（PipedInputStream）。如果要进行管道输出，就必须把输出流连到输入流上。

下面是验证管道流的示例。

```java
import java.io.* ;
public class PipedDemo{
    public static void main(String args[]){
        Send s = new Send() ;
        Receive r = new Receive() ;
        try{
            s.getPos().connect(r.getPis()) ;   // 连接管道
        }catch(IOException e){
            e.printStackTrace() ;
        }
        new Thread(s).start() ;                // 启动线程
        new Thread(r).start() ;                // 启动线程
    }
}
class Send implements Runnable{                // 线程类
    private PipedOutputStream pos = null ;     // 管道输出流
    public Send(){
        this.pos = new PipedOutputStream() ;   // 实例化输出流
    }
    public void run(){
        String str = "Welcome you!!!" ;        // 要输出的内容
        try{
            this.pos.write(str.getBytes()) ;
        }catch(IOException e){
            e.printStackTrace() ;
        }
        try{
            this.pos.close() ;
```

```java
            }catch(IOException e){
                e.printStackTrace() ;
            }
        }
        public PipedOutputStream getPos(){    // 得到此线程的管道输出流
            return this.pos ;
        }
    }
class Receive implements Runnable{
        private PipedInputStream pis = null ;    // 管道输入流
        public Receive(){
            this.pis = new PipedInputStream() ;    // 实例化输入流
        }
        public void run(){
            byte b[] = new byte[1024] ;    // 接收内容
            int len = 0 ;
            try{
                len = this.pis.read(b) ;    // 读取内容
            }catch(IOException e){
                e.printStackTrace() ;
            }
            try{
                this.pis.close() ;    // 关闭
            }catch(IOException e){
                e.printStackTrace() ;
            }
            System.out.println("接收的内容为:" + new String(b,0,len)) ;
        }
        public PipedInputStream getPis(){
            return this.pis ;
        }
    }
```

程序运行结果如下：

```
接收的内容为：Welcome you!!!
```

示例中定义了两个线程对象：在发送的线程类中定义了管道输出流，在接收的线程类中定义了管道输入流。在操作时只需要使用 PipedOutputStream 类中提供的 connect() 方法就可以将两个线程管道连接在一起，线程启动后会自动进行管道的输入/输出操作。

7.7 BufferedReader 类和 BufferedWriter 类

BufferedReader 类用于从缓冲区中读取内容，所有的输入字节数据都将放在缓冲区；BufferedWriter 类用于写入数据到缓冲区。

BufferedReader 类是 Reader 类的子类，使用该类可以以行为单位读取数据；BufferedWriter 类是 Writer 类的子类，该类可以以行为单位写入数据。例如：

```java
import java.io.BufferedReader;
import java.io.BufferedWriter;
import java.io.File;
import java.io.FileReader;
import java.io.FileWriter;
public class BufferedDemo {
    public static void main(String[] args) throws Exception{
        // 创建 BufferedReader 对象
        FileReader fr = new FileReader("d:\\example1.txt");
        File f = new File("d:\\example2.txt");
        // 创建文件输出流
        FileWriter fw = new FileWriter(f);
        BufferedReader br = new BufferedReader(fr);
        // 创建 BufferedWriter 对象
        BufferedWriter bw = new BufferedWriter(fw);
        String str = null;
        while ((str = br.readLine())!= null) {
            bw.write(str + "\n"); // 为读取的文本行添加回车
        }
        br.close();              // 关闭输入流
        bw.close();              // 关闭输出流
    }
}
```

7.8 数据操作流

Java io 包中提供了两个与平台无关的数据操作流，分别为数据输出流（DataOutputStream）和数据输入流（DataInputStream）。通常数据输出流会按照一定的格式将数据输出，再通过数据输入流按照一定的格式将数据读入，这样方便对数据进行处理。

数据输出流（DataOutputStream）和数据输入流（DataInputStream）分别有各种 writeXxx() 方法和 readXxx() 方法对数据进行输出和读取。下面的示例演示了把订单数据通过数据输出流保存到文件，然后使用数据输入流从文件中读取出来的过程。

```java
import java.io.DataOutputStream ;
import java.io.File ;
import java.io.FileOutputStream ;
public class DataOutputStreamDemo{
    public static void main(String args[]) throws Exception{
```

```java
            DataOutputStream dos = null ;        // 声明数据输出流对象
            File f = new File("d:" + File.separator + "order.txt") ;
                                                  // 实例化数据输出流对象
    dos = new DataOutputStream(new FileOutputStream(f)) ;
            String names[] = {"衬衣","手套","围巾"} ;// 商品名称
            float prices[] = {98.3f,30.3f,50.5f} ; // 商品价格
            int nums[] = {3,2,1} ;                // 商品数量
            for(int i=0;i<names.length;i++){      // 循环输出
                dos.writeChars(names[i]) ;        // 写入字符串
                dos.writeChar('\t') ;             // 写入分隔符
                dos.writeFloat(prices[i]) ;       // 写入价格
                dos.writeChar('\t') ;             // 写入分隔符
                dos.writeInt(nums[i]) ;           // 写入数量
                dos.writeChar('\n') ;             // 换行
            }
            dos.close() ;                         // 关闭输出流
    }
}
```

以上程序是将订单数据写入到 order.txt 文件中。下面的程序是从 order.txt 文件中读取数据。

```java
    import java.io.DataInputStream ;
    import java.io.File ;
    import java.io.FileInputStream ;
    public class DataInputStreamDemo{
        public static void main(String args[]) throws Exception{
DataInputStream dis = null ;                      // 声明数据输入流对象
                                                  // 文件的保存路径
    File f = new File("d:" + File.separator + "order.txt") ;
                                                  // 实例化数据输入流对象
    dis = new DataInputStream(new FileInputStream(f)) ;
            String name = null ;                  // 接收名称
            float price = 0.0f ;                  // 接收价格
            int num = 0 ;                         // 接收数量
            char temp[] = null ;                  // 接收商品名称
            int len = 0 ;                         // 保存读取数据的个数
            char c = 0 ;                          // '\u0000'
            try{
                while(true){
                    temp = new char[200] ;        // 开辟空间
                    len = 0 ;
                    while((c=dis.readChar())!='\t'){   // 接收内容
                        temp[len] = c ;
```

```
                                    len ++ ;                          // 读取长度加 1
                        }
// 将字符数组变为 String
name = new String(temp,0,len) ;
                        price = dis.readFloat() ;        // 读取价格
                        dis.readChar() ;                 // 读取 \t
                        num = dis.readInt() ;            // 读取 int
                        dis.readChar() ;                 // 读取 \n
                        System.out.printf(" 名称：%s；价格：%5.2f；数量：
                                %d\n",name,price,num) ;
                }
        }catch(Exception e){}
        dis.close() ;
    }
}
```

7.9 对象流

Java 提供了 ObjectInputStream 与 ObjectOutputStream 类读取和保存对象，它们分别是对象输入流和对象输出流。ObjectInputStream 类和 ObjectOutputStream 类是 InputStream 与 OutputStream 类的子类，继承了它们所有的方法。

下面的示例是在 D 盘存在 login.txt 文件，实现用户密码的修改。

```java
import java.io.FileInputStream;
import java.io.FileOutputStream;
import java.io.ObjectInputStream;
import java.io.ObjectOutputStream;
import java.io.Serializable;
public class ObjectDemo {
    public static void main(String[] args) throws Exception{
        User user = new User("sam", "8080");    // 创建 user 类的对象
        FileOutputStream fos=new FileOutputStream("d:\\login.txt");
        // 创建输出流对象，使之可以将对象写入文件中
        ObjectOutputStream obs = new ObjectOutputStream(fos);
        obs.writeObject(user);                   // 将对象写入文件中
        System.out.println(" 写入文件的信息 ");
        System.out.println(" 用户名: " + user.name);
        System.out.println(" 密码: " + user.password);
        FileInputStream fis = new FileInputStream("d:\\login.txt");
        // 创建输入流对象，使之可以从文件中读取数据
        ObjectInputStream ois = new ObjectInputStream(fis);
        user = (User)ois.readObject();           // 读取文件中的信息
```

```
            user.setPassword("666666");              // 修改密码
            System.out.println("修改后的文件信息");
            System.out.println("用户名: " + user.name);
            System.out.println("密码: " + user.password);
    }
}
class User implements Serializable{
    String name;
    String password;
    User(String name,String password){
            this.name = name;
            this.password = password;
    }
    public void setPassword(String password) {
            this.password = password;
    }
}
```

程序运行结果如图 7.2 所示。

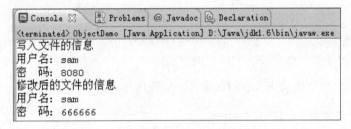

图 7.2 程序运行结果

程序中对类 User 使用了对象序列化。对象序列化可以实现直接存取对象。将对象存入一个流称为序列化，而从一个流将对象读出称为反序列化。

7.10 Scanner 类

Scanner 类是 JDK1.5 新增的类，是 java.util 包中的类。该类用于实现用户的输入，是一种只要有控制台就能实现输入操作的类。如果程序操作数据流只为读取文本数据，那么建议使用 Scanner 类实现具体操作，本书不再赘述，请读者查看 Java API 文档。

7.11 要点总结

本章针对 Java 语言的输入 / 输出技术进行了细致讲解。通过本章的讲解，读者应该熟练掌握 Java 语言中输入/输出流的操作，这里所指的流的操作包括文件输入/输出流、缓冲输入/输出流、打印输入/输出流、管道输入/输出流、数据输入/输出流和对象输入/输出流等。另外，对于数据流必须根据具体情况，有选择地使用字节流或字符流。

7.12 编程练习

1. 编写一个程序，从名为 handson.txt 的文件中读取并显示用户名和密码。如果源文件不存在，就显示相应的错误信息。

2. 编写一个程序，接收从键盘输入的数据，并把从键盘输入的内容写到 handsoninput.txt 文件中，如果输入"quit"，程序就结束。

第 8 章
Java 数据库编程

因为在数据库编程中一切都是以 SQL 语句为操作标准的,所以只有掌握 SQL 语法,才能更加方便地开发用户所需要的程序。在 JDBC 中所有的类和接口都保存在 java.sql 包中。实际上,JDBC 本身是一种操作标准,所有的数据库生产商,只要是想支持 Java,就必须符合 JDBC 规范。

8.1 JDBC 技术

JDBC 是一套允许 Java 同一个 SQL 数据库对话的程序设计接口。JDBC 经常被认为是 Java Database Connectivity 的缩写。事实上,JDBC 是一个产品的商标名,由一组用 Java 编程语言编写的类和接口组成。

8.1.1 JDBC 技术简介

JDBC 为开发人员提供了一个标准的 API,使他们能够用纯 Java API 实现数据库应用程序。试想如果没有一个标准的 API,开发者就不得不针对不同的数据库编写不同的程序代码。使用 JDBC 就不必为访问 Microsoft SQL Server 数据库写一段代码,而又为访问 Oracle 数据库专门写另一段代码。结合 Java 的平台无关性,几乎可以编写一段代码在任何一种平台下操作任何一种数据库。

JDBC 是用于将 Java 程序和关系数据连接起来的程序接口。通过这个接口,开发者将访问请求语句以 SQL 语句的形式编写出来，然后由该程序接口传送到数据库，结果再由同一个接口返回。

JDBC 在设计之初力求达到以下三个目标。

（1）低级的 API，即 SQL 级 API。

（2）JDBC 可以建立在现有的数据库访问接口上。

（3）使用简单。

JDBC 的优点在于与 ODBC 十分相似，有利于用户理解，便于进一步封装复用，加强了程序的可移植性，并提供了与 ODBC 的桥接方法。

简单来说，JDBC 可以完成以下三件事情。

（1）与特定的数据库进行连接。

（2）向数据库发送 SQL 语句，实现对数据库的特定操作。

（3）对数据库返回的结果进行处理。

它对编程人员屏蔽了很多细节上的问题，从而可以很好地简化和加快数据库程序的开发过程。

8.1.2 JDBC 驱动程序

驱动程序主要用于把应用程序的 API 转化为对特定数据库的请求。目前的 JDBC 驱动程序可分为以下 4 种类型。

1. 第一类：JDBC-ODBC 桥

JDBC-ODBC Bridge 是一种 JDBC 驱动程序，由 JavaSoft 公司提出、INTERSOIV 公司开发。它充分发挥了 ODBC 支持大量数据源的优势。JDBC 利用 JDBC-ODBC Bridge，通过 ODBC 操作数据库。因为 ODBC 早已成为数据库访问的业界标准，所以 JDBC-ODBC Bridge 的出现更多的是出于市场因素的考虑。它最直接的缺点就是依赖 ODBC，这样 ODBC 的局限性将存在于使用 JDBC-ODBC Bridge 作为驱动的程序中。

2. 第二类：本地 API 部分 Java 驱动程序

本地 API 部分 Java 驱动程序（Java to Native API）是利用客户机上的本地代码库与数据库直接通信。因为这类驱动程序必须使用本地库，所以这些库必须实现安装在客户机上。不过大多数的数据库供应商都为其产品提供了该类型的驱动程序。

3. 第三类：JDBC-Net 纯 Java 驱动程序

JDBC-Net 纯 Java 驱动程序是面向数据库中间件的纯 Java 驱动程序，JDBC 调用被转换成一种中间件厂商的协议，中间件再把这些调用转换到数据库 API。这种驱动程序比较灵活，能够发布到网上，常被用在三层网络解决方案中。这种驱动程序通常由一些与数据库产品无关的公司开发。

4. 第四类：纯 Java 的驱动程序

纯 Java 的驱动程序通过本地协议直接与数据库引擎相连接。配合合适的通信协议，这种驱动程序也可以用于 Internet。由于数据库引擎和客户机之间没有本地代码层或中间软件，因此具有明显的性能优势。

其中，第三、四类都是纯 Java 的驱动程序，对于 Java 开发者来说，它们在性能、可移植性、功能等方面都有优势。

8.1.3 JDBC 和 ODBC 与其他 API 的比较

早在 JDBC 还未成形之时，Microsoft 的 ODBC 已经成为数据库访问的业界标准，是使用非常广泛的访问关系数据库的编程接口。它几乎能在所有平台上连接所有的数据库。但 ODBC 不适合直接在 Java 中使用，因为其使用 C 语言接口。从 Java 调用本地 C 代码在安全性、实现、坚固性和程序的自动移植性方面都有许多缺点。

不妨将 JDBC 想象成被转换为面向对象接口的 ODBC，而面向对象的接口对 Java 开发者来说会更加容易理解。ODBC 把简单和高级功能混在一起，这使得理解和学习 ODBC 变得更复杂，而 JDBC 尽量保证简单功能的简便性，并在必要时允许使用高级功能。从 Java 的平台无关性来讲，也迫切需要使用纯 Java 的 API。如果使用 ODBC，就必须手动将 ODBC 驱动程序管理器和驱动程序安装在每台客户机上，而完全用 Java 编写 JDBC 驱动程序在所有 Java 平台上都可以自动安装、移植并保证安全性。

JDBC API 对于基本的 SQL 抽象和概念是一种自然的 Java 接口，建立在 ODBC 上，并保留了 ODBC 的基本设计特征。因此，熟悉 ODBC 的开发者会发现 JDBC 很容易使用。事实上，这两种接口都基于 X/Open SQL CLI（调用级接口），最大的区别在于 JDBC 以 Java 风格与优点为基础并进行优化，更易于使用。

Microsoft 随后又引入了 ODBC 之外的新 API，如 RDO、ADO 和 OLE DB。在设计方面，这些接口与 JDBC 相同，都是面向对象的数据库接口且基于可在 ODBC 上实现的类。尽管这些接口各有特点，但却不能替代 ODBC。

8.2 结构化查询语言

结构化查询语言（Structured Query Language，SQL）是一种数据库查询和程序设计语言，用于存取数据及查询、更新和管理关系数据库系统，同时也是数据库脚本文件的扩展名。

8.2.1 SQL 简介

SQL 语言是 IBM 公司在 20 世纪 70 年代所开发的一种结构化查询语言。它是一个综合的、通用的、功能强大的关系型数据库语言，能实现数据库的创建、更新、删除、数据定义、文本限制、出现控制等操作，被公认为数据库操作不可或缺的工具。

SQL 现在已经成为关系数据库的标准语言。美国国家标准协会（ANSI）和国际标准化组织（ISO）制定了一系列的 SQL 标准。

SQL 语言集数据查询、数据操纵、数据定义和数据控制功能于一体。SQL 语言具有以下特点：

（1）综合统一。

（2）高度非过程化。

（3）面向集合的操作方式。

（4）以同一种语法结构提供两种使用方式。

（5）语言简洁，易学易用。

8.2.2 SELECT 语句

SQL 语言提供了 SELECT 语句进行数据库查询。SELECT 语句的一般格式为：

```
SELECT [ALL|DISTINCT] <目标列表达式>[,<目标列表达式>]…
FROM <表名或视图名> [,<表名或视图名>]…
[WHERE<条件表达式>]
[GROUP BY<列名>[HAVING<条件表达式>]]
[ORDER BY<列名>[ASC|DESC]]
```

其含义是：根据 WHERE 子句的条件表达式，从 FROM 子句指定的基本标或视图中找到满足条件的元组，再按 SELECT 子句中的目标列表达式选出元组中的属性值形成结果表。如果有 GROUP 子句，就将结果表按列名的值进行分组，该属性列值相等的元组为一个组。通常会在每组中使用聚集函数。如果 GROUP 子句带 HAVING 短语，那么只有满足指定条件的组才能作为结果返回。如果有 ORDER 子句，结果就按列名的值升序或降序排列。

例如：有一个学生表（Student）由学号（Sno）、姓名（Sname）、性别（Ssex）、年龄（Sage）、所在系（Sdept）5 个属性组成。

```
SELECT * FROM Student;
```

将返回 Student 表所有内容。

```
SELECT Sname FROM Student;
```

将只返回 Student 表中所有学生的名字。

```
SELECT Sname FROM Student WHERE Sno = 00070713;
```

将只返回 Student 表中学号为 00070713 学生的名字。

```
SELECT Sname FROM Student ORDER BY Sno;
```

将只返回 Student 表中所有学生的名字，并按学号的升序排列。

8.2.3 更新记录

SQL 中数据更新包括插入数据、修改数据和删除数据。

1. INSERT

SQL 的数据插入语句 INSERT 通常有两种形式：一种是插入单条数据；另一种是插入子查询结果。限于本书篇幅，这里只介绍插入单条数据的 INSERT 语句的格式。插入单条记录的 INSERT 语句格式如下：

```
INSERT
INTO< 表名 >[(< 属性列 1>[,< 属性列 2>…])]
VALUES(< 常量 1>[,< 常量 2>]…);
```

其含义是将新数据插入指定表中。

例如：将一个新的学生记录插入到上一小节提到的 Student 表中。

```
INSERT INTO Student VALUES (00070713,'Zidane','Male',34,'Real Madrid');
```

2. UPDATE

SQL 语言提供了 UPDATE 语句进行数据修改。UPDATE 语句的一般格式如下：

```
UPDATE < 表名 >
SET  < 列名 >=< 表达式 >[,< 列名 >=< 表达式 >]…
[WHERE< 条件 >];
```

其含义是修改指定表中满足 WHERE 子句条件的数据。其中 SET 子句给出 < 表达式 > 的值用于取代相应的属性列值。如果省略 WHERE 子句，就表示要修改表中所有的数据。

例如：在上一小节提到的 Student 表中，将学号为 00070713 的学生年龄改为 35 岁。

```
UPDATE Student SET Sage = 35 WHERE Sno=00070713;
```

3. DELETE

SQL 语言提供了 DELETE 语句进行数据删除。DELETE 语句的一般格式如下：

```
DELETE
FROM< 表名 >
[WHERE < 条件 >];
```

其含义是从指定表中删除满足 WHERE 子句条件的所有数据。如果省略 WHERE 子句，就表示删除表中全部数据，但表仍然存在。

例如：删除上一小节提到的 Student 表中年龄大于 18 岁的学生记录。

```
DELETE FROM Student WHERE Sage>18;
```

8.2.4 聚集函数

SQL 中提供了 5 种聚集函数，分别用于统计记录数目、平均值、最大值、最小值和求和。

1. COUNT

函数 COUNT() 用于统计记录数目。

例如：

```
SELECT COUNT（*） FROM Student;
```

将返回 Student 表中记录的数目。

2. AVG

函数 AVG() 用于计算返回结果中特定字段的平均值。

例如：

```
SELECT AVG（Sage） FROM Student;
```

将返回 Student 表中记录的所有学生的平均年龄。

3. MAX

函数 MAX() 用于计算返回结果中特定字段的最大值。

例如：

```
SELECT MAX（Sage） FROM Student;
```

将返回 Student 表中记录的所有学生中最大年龄。

4. MIN

函数 MIN() 用于计算返回结果中特定字段的最小值。

例如：

```
SELECT MIN（Sage） FROM Student;
```

将返回 Student 表中记录的所有学生中最小年龄。

5. SUM

函数 SUM() 用于计算返回结果中特定字段值的总和。

例如：

```
SELECT SUM（Sage） FROM Student;
```

将返回 Student 表中记录的所有学生的年龄总和。

8.3 JDBC 基本操作

在 JDBC 的基本操作中，常用的类和接口是 DriverManager、Connection、Statement、ResultSet、PreparedStatemen，都放在 java.sql 包中，相关功能如表 8-1 所示。

表8-1 java.sql包中数据库操作的接口和类

序号	类/接口	功 能 说 明
1	DriverManager类	用于加载和卸载各种驱动程序并建立与数据库的连接
2	Connection接口	此接口表示与数据的连接
3	Statement接口	此接口用于执行 SQL 语句并将数据检索到 ResultSet 中
4	ResultSet接口	此接口表示查询出来的数据库数据结果集
5	PreparedStatemen接口	此接口用于执行预编译的 SQL 语句
6	Date类	包含将 SQL 日期格式转换为 Java 日期格式的各种方法

8.3.1 JDBC 操作步骤

数据库安装并配置完成后，即可按如图 8-1 所示步骤进行数据库的操作。

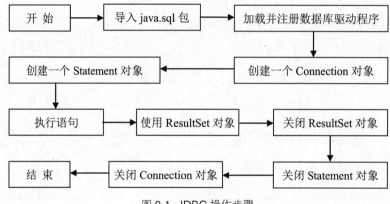

图 8-1 JDBC 操作步骤

（1）加载数据库驱动程序：各个数据库都会提供 JDBC 的驱动程序开发包，直接把 JDBC 操作所需要的开发包（一般为 *.jar 或 *.zip）配置到项目的 Libraries 中。

（2）连接数据库：根据各个数据库的不同，连接的地址也不同，此地址将由数据库厂商提供。一般在使用 JDBC 连接数据库时，都要求用户输入数据库连接的用户名和密码，如本章使用的数据库是 SQL Server 2000，用户名为 sa，密码为空，用户在取得连接后才可以对数据库进行查询或更新的操作。

（3）使用语句进行数据库操作：数据库操作分为更新和查询两种，除了可以使用标准的 SQL 语句外，对于各个数据库也可以使用其自己提供的各种命令。

（4）关闭数据库连接：数据库操作完毕后，依次关闭连接以释放资源。

8.3.2 JDBC-ODBC 连接数据库

JDBC-ODBC 连接数据库称为桥连接。首先将 JDBC 首先翻译为 ODBC，然后使用 ODBC 驱动程序与数据库通信，必须安装 ODBC 驱动程序和配置 ODBC 数据源，本章所用数据库是 SQL Server 2000 的 userDB 数据库，如图 8-2 所示。

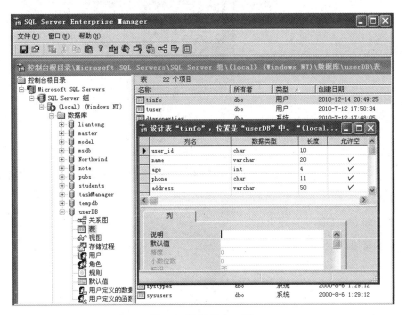

图 8-2 数据库 userDB

配置 ODBC 数据源的步骤如下。

步骤 01 执行【控制面板】→【管理工具】→【数据源(ODBC)】命令，打开如图 8-3 所示的对话框。

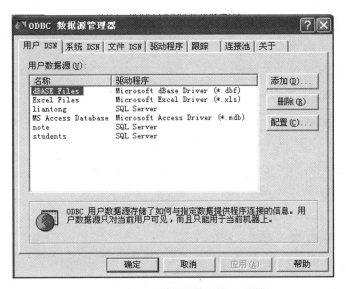

图 8-3 "ODBC 数据源管理器"对话框

零基础轻松学 Java

步骤 02 单击【添加】按钮，打开【创建数据源】对话框，选择数据库的驱动程序，如图 8-4 所示。

图 8-4 "创建新数据源"对话框

步骤 03 单击【完成】按钮，如图 8-5 所示的对话框，在此命名数据源，如 userDB，描述此数据源的数据库是"用户信息库"，服务器选择本地，输入·或 local。

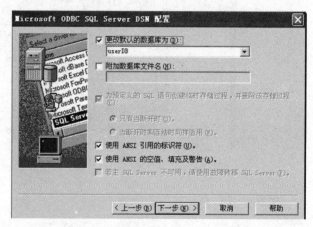

图 8-5 "Microsoft ODBC SQL Server DSN 配置"对话框

步骤 04 单击两次【下一步】按钮，打开"Microsoft ODBC SQL Server DSN 配置"对话框，设置"更改默认数据库为："为 userDB，单击【下一步】→【完成】按钮，进入如图 8-6 所示的"ODBC Microsoft SQL Sevrer 安装"对话框，单击"测试数据源"按钮，打开如图 8-7 所示的对话框，提示测试成功。此时在图 8-3 中的用户数据源"名称"中会出现刚刚配置成功的数据源的相关信息。

图 8-6 "ODBC Microsoft SQL Sevrer 安装"对话框

图 8-7 "SQL Sevrer ODBC 数据源测试"对话框

下面示例是连接上面配置的数据源 userDB 的代码，请读者掌握连接步骤。

```java
import java.sql.Connection;
import java.sql.DriverManager;
import java.sql.ResultSet;
import java.sql.SQLException;
import java.sql.Statement;
public class Test_JDBC01 {
    public static final String DBDRIVER =
        "sun.jdbc.odbc.JdbcOdbcDriver";              // 定义数据库驱动程序
    // 定义数据库的连接地址,userDB 为配置成功的数据源名称
    public static final String DBURL = "jdbc:odbc:userDB";
    // 连接数据库的登录用户名、密码
    public static final String DBUSER = "sa";
    public static final String PASSWORD = "";
    public static void main(String[] args) {
        onnection conn = null;                       // 创建数据库连接对象
        Statement stmt = null;      // 定义 Statement 对象，用于操作数据库
        ResultSet rs = null;                         // 创建数据库结果集对象
        String sql = "select * from tinfo";          // 数据库查询语句字符串
        // 1.注册数据库驱动程序
        try {
            Class.forName(DBRIVER);
        } catch (ClassNotFoundException e) {
            System.out.println("加载驱动程序失败！请检查！");
            e.printStackTrace();
        }
        // 2.获取数据库的连接
        try {
```

```java
            conn=DriverManager.getConnection(DBURL,DBUSER,PASSWORD);
        } catch (SQLException e) {
            System.out.println("连接数据库失败，请检查用户名和密码！");
            e.printStackTrace();
        }
        // 3.获取表达式（根据连接创建语句对象）
        try {
            stmt = conn.createStatement();
        } catch (SQLException e) {
            System.out.println("获取表达式出错！");
            e.printStackTrace();
        }
        //4.执行SQL语句
        try {
            rs = stmt.executeQuery(sql);
        } catch (SQLException e) {
            System.out.println("执行SQL语句出错！");
            e.printStackTrace();
        }
        //5.显示结果集数据
        try {
            while(rs.next()){// 像游标在第一记录的上面
                System.out.print(rs.getString("user_id")+"\t");
                System.out.print(rs.getString(2)+"\t");
                System.out.print(rs.getInt(3)+"\t");
                System.out.print(rs.getString(4)+"\t");
                System.out.println(rs.getString(5));
            }
        } catch (SQLException e) {
            e.printStackTrace();
        }
        //6.释放资源
        try {
            rs.close();
            stmt.close();
            conn.close();
        } catch (SQLException e) {
            System.out.println("释放资源失败！");
            e.printStackTrace();
        }
    }
}
```

运行结果如下：

20100801	佟麟	20	13033939898	电厂车间主任
20100802	毛平	18	13679802322	移动公司业务经理
20100803	郝蕾	30	15878749282	电影公司司仪
20080804	张永刚	19	15098023056	组织部部长

数据库的驱动程序 DBDRIVER = "sun.jdbc.odbc.JdbcOdbcDriver" 是 eclipse 所建项目 JRE System Liberaryrt.jar 包中的一个类 JdbcOdbcDriver.class，所在的包为 sun.jdbc.odbc，如图 8-8 所示。

图 8-8 驱动程序类所在 jar 包

因为此类已经导入到项目的运行环境库中，所以可以直接使用。

1. 加载 JDBC 驱动

在通过 JDBC 与数据库建立连接之前，必须加载相应数据库的 JDBC 驱动。调用方法 Class.forName() 将显式地将驱动程序添加到 java.lang.System 的属性 jdbc.drivers 中。

如示例中语句：

```
Class.forName(DBDRIVER);
```

第一次调用 DriverManager 类的方法时，将自动加载这些驱动程序类。DriverManager 类将搜索系统属性 jdbc.drivers，如果用户已输入一个或多个驱动程序，那么 DriverManager 类将试图加载它们。

也可以用其他方法加载，如 new sun.jdbc.odbc.JdbcOdbcDriver();，建议使用示例的方法。

 Class.forName() 方法中使用的驱动程序类名由驱动发布者提供，如下面要讲的连接方式。同时还要保证 JDBC 驱动在构建路径下。

2. Connection 接口

Connection 接口代表与数据库的连接。连接过程包括所执行的 SQL 语句和在该连接上所返回的结果。可以通过调用 DriverManager.getConnection() 方法实现与数据库建立连接。

```
public static Connection getConnection(String url,String user,String
password) throws SQLException
```

其参数含义如下：

- url：指 JDBC URL，提供了一种标识数据库的方法，可以使相应的驱动程序识别该数据库并与之建立连接。URL 的标准语法如下：

```
jdbc:< 子协议 >:< 子名称 >
```

如示例中语句：

```
public static final String DBURL = "jdbc:odbc:userDB";
conn=DriverManager.getConnection(DBURL,DBUSER,PASSWORD);
```

 开发者可以参考驱动程序的相关说明文档获得正确的 url 拼写方式。

- DBUSER：连接数据库的用户名。
- PASSWORD：连接数据库的密码。

连接建立之后，就可以用来向它所连接的数据库传送 SQL 语句了。JDBC 提供了三个接口：Statement、PreparedStatement 和 CallableStatement，用于向数据库发送 SQL 语句。Connection 接口中定义的方法用于返回这三个接口。

返回 Statement 接口的常用方法如下：

```
Statement createStatement() throws SQLException
```

返回 PreparedStatement 接口的常用方法如下：

```
PreparedStatement prepareStatement(String sql) throws SQLException
```

返回 CallableStatement 接口的常用方法如下：

```
CallableStatement prepareCall(String sql) throws SQLException
```

对于初学者来说，CallableStatement 并不常用。

需要注意的是，在确定使用完 Connection 接口后，应该调用其 close 方法断开连接。

 初学者经常忘记调用 close 方法断开连接。根据操作步骤，是放在最后断开的。

3. Statement 接口

Statement 接口用于将 SQL 语句发送到所连接的数据库中。创建 Statement 接口后就可以使用它执行 SQL 语句了。

Statement 接口提供了三种执行 SQL 语句的方法：executeQuery、executeUpdate 和 execute。开发者应根据这三种方法的适用范围选择使用。

（1）executeQuery 方法用于产生单个结果集的语句，如 SELECT 语句，示例中查询 tinfo 表中的记录。

（2）executeUpdate 方法用于执行 INSERT、UPDATE、DELETE 及 SQL DDL（数据定义语言）语句，如 CREATE TABLE 和 DROP TABLE。executeUpdate 的返回值是一个整数，表示执行 SQL 语句后受影响的记录数。

（3）execute 方法用于执行返回多个结果集、多个更新计数或二者组合的语句。

在 Statement 接口使用结束后，应该调用其 close 方法关闭。

 与 Connection 类似，初学者经常忘记调用 close 方法关闭 Statement。

4. ResultSet 接口

ResultSet 接口包含符合 SQL 语句中条件的所有行，其中有查询所返回的列标题及相应的值；通过一系列 get 方法访问这些行中的数据。ResultSet 中维持了一个指向当前行的指针，这个指针最初指向表的第一行之前。ResultSet.next 方法用于移动到 ResultSet 中的下一行，使下一行成为当前行。ResultSet.next 方法返回一个 boolean 类型的值，如果这个值是 True，就说明已经成功地移动到下一行；如果这个值是 False，就说明表已经到了最后一行。

在每一行内，可按任何次序获取列值。但为了保证可移植性，应该从左至右获取列值，并一次性地读取列值。列名或列号可用于标识要从中获取数据的列。例如，如果 ResultSet 对象 rs 的第一列列名为"user_id"，并将值存储为字符串，就可以通过以下两种方式访问该列的值。

```
rs.getString("user_id");
rs.getString(1);
```

 列号是从左至右编号的，并且从 1 开始而不是 0。

对于一系列 get 方法，JDBC 驱动程序试图将基本数据转换为指定的 Java 类型，然后返回适合的 Java 值。例如，如果 getXXX 方法为 getString，而基本数据库中数据类型为 VARCHAR，那么 JDBC 驱动程序将把 VARCHAR 转换为 Java 中的 String 类型。不再使用 ResultSet 时，应调用其 close 方法关闭。

 ResultSet 使用完毕后也要调用其 close 方法关闭，然后关闭 Statement，最后断开数据库连接。

8.3.3 JDBC 直接连接数据库

JDBC 桥连的方式虽然简单，但是需要安装 ODBC 驱动程序和配置 ODBC 数据源，得到 ODBC 的支持。可以直接使用该厂商提供的驱动程序与数据库进行连接。

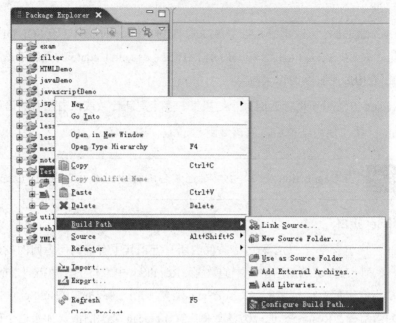

图 8-9 Configure Build Path

要求连接之前将厂商提供的驱动程序的 jar 包或 zip 包导入到项目的 Libraries 中，如图 8-9 所示。单击 Add JARs 按钮，找到驱动程序的 jar 包所在的目录，如图 8-10 所示。

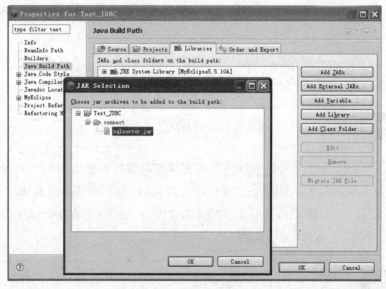

图 8-10 加载驱动程序 jar 包

单击 OK 按钮，关闭 JAR Selection 对话框，再单击 OK 按钮，完成配置。此时，Microsoft SQL Server 2000 的驱动程序已经加载到了项目中，如图 8-11 所示。

第 8 章　Java 数据库编程

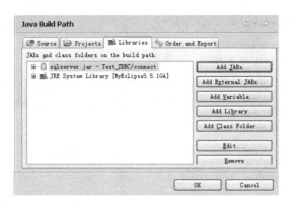

图 8-11　Java 项目的配置路径

在 Eclipse 中打开如图 8-12 所示的数据库管理器 My Eclipse Database Explorer。

在 DB Browser 中右击数据库，选择 Edit 选项，如图 8-13 所示。

图 8-12　数据库管理器　　　　　图 8-13　编辑数据库连接驱动器

打开 Database Driver 对话框，如图 8-14 所示。

- Driver template：选择 Microsoft SQL Server 2000。
- Driver name：用户自定义名字，图示为 sqlserver2000。
- Connection URL：图示为 jdbc:microsoft:sqlserver://localhost:1433;DatabaseName=user DB。

不同的数据库，URL 也不相同。

```
User name: sa;
```

- Password：这里为空，如果 Microsoft SQL Server 2000 的登录密码不为空，就必须填写。
- Driver JARs：选择驱动程序所在的路径。单击 Add JARs 按钮，选择驱动程序所在的路径，则自动在左边出现。
- Driver classname：当添加驱动程序 jar 包成功后，会自动出现驱动程序的类名。单击 Finish 按钮完成配置，如图 8-15 所示。

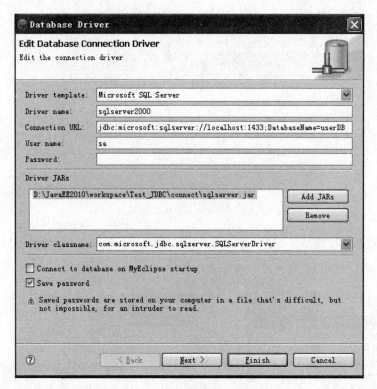

图 8-14 编辑数据库连接驱动

在数据库管理器的 DB Browser 中单击连接图标（带箭头的图标）查看是否连接成功。如果没有连接成功，请检查数据库配置尝试是否正确，或者查看 Microsoft SQL Server 2000 是否启动。连接成功后，数据库操作所需的驱动程序名、URL、用户名和密码直接复制到代码中。

图 8-15 数据库连接成功

下面的示例演示了直连数据库的操作。

```java
public class Test_JDBC02 {
    private Connection conn = null;
    private Statement stmt = null;
    private ResultSet rs = null;
```

```java
        private final static String DRIVER = 
                            "com.microsoft.jdbc.sqlserver.SQLServerDriver";
        private final static String URL = 
        "jdbc:microsoft:sqlserver://localhost:1433;DatabaseName=userDB";
        private final static String USER = "sa";
        private final static String PASSWORD = "";
        public Connection getConn(){
            try {
                Class.forName(DRIVER);
                conn = DriverManager.getConnection(URL, USER, PASSWORD);
            } catch (ClassNotFoundException e) {
                e.printStackTrace();
            } catch (SQLException e) {
                e.printStackTrace();
            }
            if (conn!=null) {
                return conn;
            } else {
                System.out.println(" 数据库连接失败,请检查! ");
                return conn;
            }
        }
        public void release(){
            try {
                rs.close();
                stmt.close();
                conn.close();
            } catch (SQLException e) {
                e.printStackTrace();
            }
        }
        public static void main(String[] args) {
            Test_JDBC02 test = new Test_JDBC02();
            test.getConn();
            test.release();
        }
    }
```

8.3.4 JDBC 对数据库的更新操作

数据库连接完成后,就可以对数据库进行操作了。JDBC 对数据库的查询操作比较简单,前面的示例已经讲解过,下面使用 Statement 接口分别完成数据库的插入、修改、删除操作。

1. 数据库的插入操作

下面向 userDB 数据库的 tinfo 表中增加一条新的记录，并通过 Statement 接口执行。为了读者观看方便，将所有的异常直接在主方法抛出以减少程序中的 try…catch 代码。

```java
import java.sql.*;
public class InsertDemo01 {
    private final static String DRIVER =
                    "com.microsoft.jdbc.sqlserver.SQLServerDriver";
    private final static String URL = "jdbc:microsoft:sqlserver://localhost:1433;DatabaseName=userDB";
    private final static String USER = "sa";
    private final static String PASSWORD = "";
    public static void main(String[] args) throws Exception{
        Connection conn = null;
        Statement stmt = null;
        String sql = "INSERT INTO tinfo(user_id,name,age,phone,address)" + "VALUES('20100806','洪七公',33,'8888888','丐帮帮主四海为家')";
        Class.forName(DRIVER);
        conn = DriverManager.getConnection(URL, USER, PASSWORD);
        stmt = conn.createStatement();
        stmt.executeUpdate(sql);           // 执行数据库更新操作
        stmt.close();
        conn.close();
    }
}
```

如果在插入数据时使用变量，就可以使用下面的代码编写。

```java
String user_id = "20100806";
String name = "洪七公";
int age = 33;
String phone = "8888888";
String address = "丐帮帮主四海为家";
String sql = "INSERT INTO tinfo(user_id,name,age,phone,address)" + "VALUES('"+user_id+"','"+name+"','"+age+"','"+phone+"','"+address+"')";
```

从上面的程序代码中可以发现，SQL 语句采用了拼凑的形式，实际上是多个字符串的连接，这种形式容易出错，后面讲解的 PreparedStatement 接口可以解决这个问题。

2. 数据库的修改操作

执行数据库的修改操作，只需要将 SQL 语句修改为 UPDATE 语句即可。例如：

```java
import java.sql.*;
```

```java
public class UpdateDemo01 {
    private final static String DRIVER =
                        "com.microsoft.jdbc.sqlserver.SQLServerDriver";
    private final static String URL = "jdbc:microsoft:sqlserver://localhost:1433;DatabaseName=userDB";
    private final static String USER = "sa";
    private final static String PASSWORD = "";
    public static void main(String[] args) throws Exception{
        Connection conn = null;
        Statement stmt = null;
        String user_id = "20100806";
        String name = " 郭靖 ";
        int age = 23;
        String phone = "136363633";
        String address = " 桃花岛 ";
        String sql = "UPDATE tinfo SET name='"+name+"',age="+age
+",phone='"+phone+"',address='"+address+"'" +"WHERE 
                        user_id='"+user_id+"'";
        Class.forName(DRIVER);
        conn = DriverManager.getConnection(URL, USER, PASSWORD);
        stmt = conn.createStatement();
        stmt.executeUpdate(sql);            // 执行数据库更新操作
        stmt.close();
        conn.close();
    }
}
```

3. 数据库的删除操作

与之前一样，直接执行 DELETE 的 SQL 语句即可完成记录的删除操作。例如：

```java
import java.sql.*;
public class DeleteDemo01 {
    private final static String DRIVER =
                    "com.microsoft.jdbc.sqlserver.SQLServerDriver";
    private final static String URL = "jdbc:microsoft:sqlserver://localhost:1433;DatabaseName=userDB";
    private final static String USER = "sa";
    private final static String PASSWORD = "";
    public static void main(String[] args) throws Exception{
        Connection conn = null;
        Statement stmt = null;
        String user_id = "20100806";
        String sql = "DELETE FROM tinfo WHERE user_id="+user_id;
```

```
        Class.forName(DRIVER);
        conn = DriverManager.getConnection(URL, USER, PASSWORD);
        stmt = conn.createStatement();
        stmt.executeUpdate(sql);          // 执行数据库更新操作
        stmt.close();
        conn.close();
    }
}
```

8.4 JDBC 高级操作

本节介绍 JDBC 高级操作，主要包括使用 PreparedStatemen 接口进行预处理操作和使用 CallableStatement 接口调用数据中的存储过程。

8.4.1 PreparedStatemen 接口

PreparedStatement 接口继承自 Statement 接口，继承了 Statement 的所有功能。除此之外，PreparedStatement 接口还具有一些 Statement 接口没有的特点。

PreparedStatement 属于预处理操作，在操作时，是先在数据表中准备好一条 SQL 语句，但此 SQL 语句的具体内容暂时不设置，而是之后再进行设置。

由于 PreparedStatement 对象已预编译过，因此其执行速度要高于 Statement 对象。为了提高效率，通常对于需要多次执行的 SQL 语句经常使用 PreparedStatemen 对象操作，以提高效率。

下面的示例演示了使用 PreparedStatement 完成数据的插入操作。

```
    import java.sql.*;
    public class PreparedStatementDemo01 {
    private final static String DRIVER = "com.microsoft.jdbc.sqlserver.SQLServerDriver";
    private final static String URL = "jdbc:microsoft:sqlserver://localhost:1433;DatabaseName=userDB";
        private final static String USER = "sa";
        private final static String PASSWORD = "";
        public static void main(String[] args) throws Exception{
            Connection conn = null;
            PreparedStatement pstmt = null;
            String user_id = "080601";
            String name = "陈占伟";
            int age = 32;
            String phone = "13703946666";
```

```
            String address = "计算机科学系";
            String sql = "INSERT INTO tinfo(user_id,name,age,phone,address)" + "VALUES(?,?,?,?,?)";
            Class.forName(DRIVER);
            conn = DriverManager.getConnection(URL, USER, PASSWORD);
            pstmt = conn.prepareStatement(sql);// 实例化 PreparedStatement
            pstmt.setString(1, user_id);
            pstmt.setString(2, name);
            pstmt.setInt(3, age);
            pstmt.setString(4, phone);
            pstmt.setString(5, address);
            pstmt.executeUpdate();              // 执行数据库更新操作，不需要 SQL
            pstmt.close();
            conn.close();
    }
}
```

从程序代码中可以发现，预处理就是使用"？"进行占位，每一个"？"对应一个具体的字段，在设置时，按照"？"的顺序设置即可。

下面的示例演示了使用 PreparedStatement 进行模糊查询记录的方法。查询"name"字段或"address"字段里包含"陈"的记录。

```
import java.sql.*;
public class PreparedStatementDemo02 {
private final static String DRIVER = "com.microsoft.jdbc.sqlserver.SQLServerDriver";
private final static String URL = "jdbc:microsoft:sqlserver://localhost:1433;DatabaseName=userDB";
    private final static String USER = "sa";
    private final static String PASSWORD = "";
    public static void main(String[] args) throws Exception{
        Connection conn = null;
        PreparedStatement pstmt = null;
        ResultSet rs = null;
        String keyword = "陈";
        String sql = "SELECT user_id,name,age,phone,address"+"FROM tinfo WHERE name LIKE ? OR address LIKE ?";
        Class.forName(DRIVER);
        conn = DriverManager.getConnection(URL, USER, PASSWORD);
        pstmt = conn.prepareStatement(sql);// 实例化 PreparedStatement
        pstmt.setString(1, "%"+keyword+"%");
        pstmt.setString(2, "%"+keyword+"%");
        rs = pstmt.executeQuery();
        while (rs.next()) {
            String user_id = rs.getString(1);   // 取得 user_id 内容
```

```
                    String name = rs.getString(2);       // 取得 name 内容
                    int age = rs.getInt(3);              // 取得 age 内容
                    String phone = rs.getString(4);      // 取得 phone 内容
                    String address = rs.getString(5);    // 取得 address 内容
                    // 输出满足条件的记录
                    System.out.print(user_id+"\t");
                    System.out.print(name+"\t");
                    System.out.print(age+"\t");
                    System.out.print(phone+»\t»);
                    System.out.println(address);
                }
                pstmt.close();
                conn.close();
            }
        }
```

 开发中建议使用 PreparedStatement 完成操作,既可以避免 Statement 拼凑 SQL 语句,也可以避免因输入非法字符而造成程序出错,引起系统的安全漏洞。

8.4.2 CallableStatement 接口

CallableStatement 接口主要调用数据库中的存储过程。因为 CallableStatement 接口是 PreparedStatement 接口的子接口,所以继承了 PreparedStatement 接口中的方法。

CallableStatement 对象仍然需要使用 Connection 对象来创建,创建的方法为 prepareCall(),用于执行存储过程。在使用存储过程中,可能需要传入相应的参数或得到定的结果,这里要传入 IN 或 OUT 参数。

根据存储过程中的参数不同,CallableStatement 对象的创建形式有三种。

(1)不带参数的存储过程。

例如:CallableStatement cs = conn.preparedCall("{call 存储过程名()}");

(2)传入 IN 参数的存储过程。

例如:CallableStatement cs = conn.preparedCall("{call 存储过程名(?,?,?)}");
需要使用 setXXX() 方法为相应的占位符赋值。

(3)传入 IN 或 OUT 参数的存储过程。

例如:CallableStatement cs = conn.preparedCall("{?=call 存储过程名(?,?,?)}");
需要使用 getXXX() 方法获得输出参数,setXXX() 方法为相应的占位符赋值。

8.4.3 事务处理

所谓事务是用户定义的一个数据库操作序列，这些操作要么全做，要么全不做，是一个不可分割的工作单位。在关系数据库中，一个事务可以是一条 SQL 语句、一组 SQL 语句或整个程序。事务具有 4 个特性：原子性、一致性、隔离性和持续性。

一个典型的例子是用户从 ATM 机上取款的过程，取款和划账要么全做，要么全不做。数据库通过 commit 命令保证全做，如果出现异常，就通过 rollback 命令保证全不做。

在 JDBC 中默认是自动提交的。自动提交的含义是把每一条 SQL 语句作为一个事务，而更多的时候事务是指一组 SQL 语句。在 JDBC 中，如果想进行事务处理，也需要按照指定的步骤完成。

（1）取消 Connection 中设置的自动提交方式 "conn.setAutoCommit(false);"。

（2）如果批处理操作成功，就执行提交事务 "conn.commit();"。

（3）如果操作失败，就会引发异常，在异常处理中让事务回滚 "conn.rollback();"。

8.5 要点总结

本章首先介绍了 JDBC 的相关背景知识及技术规范；然后概要介绍了结构化查询语言中常用的语句和函数，重点讲解 JDBC 的基本操作，包括各种常用接口的使用方法并举例说明；最后介绍了 JDBC 的部分高级操作。

8.6 练习题

1. 填空题

（1） JDBC 为开发人员提供了一个标准的 API，它由一组用 Java 编程语言编写的类和 _____ 组成。

（2） SQL 语言之所以能成为国际标准，是因为其集 _____、_____、_____ 和 _____ 功能于一体。

（3） SQL 中提供了 5 种聚集函数，分别是 _____、_____、_____、_____ 和 _____。

（4）调用方法 Class.forName() 将显式地将驱动程序添加到 _____ 的属性 jdbc.drivers 中。

（5）事务具有 4 个特性：原子性、_____、隔离性和 _____。

2. 选择题

（1）下列不属于 SQL 聚集函数的是 _____。

    A. SUM   B. AVG   C. COUNT   D. NVL

（2）Statement 接口中的哪个方法可以用于执行数据定义语言 _____。

    A. execute   B. addBatch   C. executeUpdate   D. executeQuery

（3）Connection 接口中的哪个方法用于获取 DatabaseMetaData 接口 _____。

    A. getMetaData   B. createStatement   C. prepareStatement   D. prepareCall

（4）下列哪个接口用于获取元数据 _____。

    A. Statement   B. PreparedStatement   C. Connection   D. DatabaseMetaData

（5）Connection 接口中的哪个方法用于设置事务自动提交 _____。

    A. commit   B. setAutoCommit   C. getAutoCommit   D. rollback

3. 问答题

（1）列举并说明 JDBC 驱动的 4 种类型。

（2）简述 JDBC 和 ODBC 的关系和异同。

8.7 编程练习

编写一个 java 应用程序，添加、修改和删除 Student 表中的记录。Student 表的结构如表 8-2 所示。

表8-2 Student表的结构

列名称	数据类型
Name	Varcahr（20）
Rollno	Numeric
Course	Varchar（20）

第 9 章
Java 网络编程

与网络编程有关的基本 API 位于 Java.NET 包中,其中包含基本的网络编程实现,该包是网络编程的基础。Java.NET 包既包含基本的网络编程类,也包含封装后专门处理 WEB 相关的处理类。本章将从网络基础开始讲起,介绍 UDP、TCP、HTTP 等相关网络程序。

9.1 网络基础

要开发网络应用程序,就必须对网络的基础知识有一定的了解。Java 的网络通信可以使用 TCP、IP、UDP 等协议。网络程序设计就是开发为用户提供网络服务的实用程序,如网络通信、股票行情、新闻资讯等。另外,网络程序设计也是游戏开发的必修课。

9.1.1 TCP/IP 网络模型

TCP/IP 起源于美国国防部高级研究规划署(DARPA)的一项研究计划——实现若干台主机的相互通信。现在 TCP/IP 已成为 Internet 上通信的标准。

与 OSI 参考模型不同,TCP/IP 参考模型只有 4 层,从下向上依次是网络接口层、网际层、传输层和应用层。

- 网络接口层:包括用于协作 IP 数据在已有网络介质上传输的协议。实际上,TCP/IP 标准并不定义与 OSI 数据链路层和物理层相对应的功能。相反,它定义像地址解析

（Address Resolution Protocol,ARP）这样的协议，提供 TCP/IP 协议数据结构和实际物理硬件之间的接口。

- 网际层：对应于 OSI 七层参考模型的网络层。本层包含 IP 协议、RIP 协议（Routing Information Protocol，路由信息协议），负责数据的包装、寻址和路由。同时还包含网间控制报文协议（Internet Control Message Protocol,ICMP）用于提供网络诊断信息。
- 传输层：对应于 OSI 七层参考模型的传输层，提供两种端到端的通信服务。其中 TCP 协议（Transmission Control Protocol）提供可靠的数据流运输服务，UDP 协议（Use Datagram Protocol）提供不可靠的用户数据包服务。
- 应用层：对应于 OSI 七层参考模型的应用层和表达层。因特网的应用层协议包括 Finger、Whois、FTP（文件传输协议）、Gopher、HTTP（超文本传输协议）、Telent（远程终端协议）、SMTP（简单邮件传送协议）、IRC（因特网中继会话）和 NNTP（网络新闻传输协议）等。

9.1.2 IP 地址与 InetAddress 类

互联网上的每一台计算机都有一个唯一的表示自己的标记，这个标记就是 IP 地址。在 Windows 操作系统中，用户可以通过执行【网上邻居】à【属性】à【Internet 协议（TCP/IP）】命令设置每一台计算机的 IP 地址。

IP 地址使用 32 位长度二进制数据表示，大部分 IP 地址都是以十进制的数据形式表示的，如 192.168.12.3。

IP 地址分类中 127.X.X.X 是保留地址，用作循环测试，在开发中经常使用 127.0.0.1 表示本机的 IP 地址。

IP 地址有 IPv4 和 IPv6 两类，IPv4 是互联网协议的第 4 个版本，也是使用较广泛的版本。但是 IPv4 已经无法满足当今互联网上的主机数量，所以在此基础上又产生了新的版本——IPv6。IPv6 可以比 IPv4 容纳更多的主机。

InetAddress 类主要表示 IP 地址，其包含两个子类：Inet4Address 和 Inet6Address，分别表示 IPv4 和 IPv6。在 Java 中提供了专门的网络开发程序包—java.net，InetAddress 类就是其中的类。

9.1.3 套接字

套接字是通信的基石，是支持 TCP/IP 协议网络通信的基本操作单元。可以将套接字看作不同主机间的进程进行双向通信的端点，其构成了单个主机内及整个网络间的编程界面。套接字存在于通信域中，通信域是为了处理一般的线程通过套接字通信而引进的一种抽象概念。套接字通常与同一个域中的套接字交换数据。

套接字之间的连接过程可以分为三个步骤：服务器监听、客户端请求和连接确认。服务器监听是指服务器端套接字并不定位具体的客户端套接字，而是处于等待连接的状态，实时监控网络状态。客户端请求是指由客户端的套接字提出连接请求,要连接的目标是服务器端的套接字。连接确认是指当服务器端套接字监听到或接收到客户端套接字的连接请求,就响应客户端套接字的请求,建立一个新的线程,把服务器端套接字的描述发给客户端,一旦客户端确认了此描述,连接就建立好了。而服务器端套接字继续处于监听状态,接收其他客户端套接字的连接请求。

 理解客户/服务器套接字之间的连接过程,对学习 Java 网络编程十分重要。

9.2 UDP 协议网络程序

9.2.1 概述

UDP（User Datagram Protocol）是一种无连接的协议,每个数据包都是一个独立的信息,包括完整的源地址或目的地址。由于 UDP 在网络上是以任何可能的路径传往目的地的,因此能否到达目的地、到达目的地的时间及内容的正确性都是不能被保证的。使用 UDP 时,每个数据包中都给出了完整的地址信息,无须再建立发送方和接收方的连接。但使用 UDP 传输数据是有大小限制的,每个被传输的数据包必须限定在 64 KB 之内。UDP 是一个不可靠的协议,发送方所发送的数据包不一定以相同的次序到达接收方。在 java.net 包中提供了两个类：DatagramSocket 和 DatagramPacket,用于支持数据包通信。DatagramSocket 类用于在程序之间建立传送数据包的通信连接,DatagramPacket 类用于表示一个数据包。

9.2.2 DatagramPacket 类

DatagramPacket 类的主要构造方法如下：

```
public DatagramPacket(byte[] buf,int length)
public DatagramPacket(byte[] buf,int length,InetAddress address,int port)
public DatagramPacket(byte[] buf,int offset,int length)
public DatagramPacket(byte[] buf,int offset,int length,InetAddress address,int port)
public DatagramPacket(byte[] buf,int offset,int length,SocketAddress address) throws SocketException
```

DatagramPacket 类的常用方法如下：

```
public InetAddress getAddress()
```

说明：返回发送或接收数据包的主机地址。

```
public byte[] getData()
```

说明：返回数据包内容。

```
public int getLength()
```

说明：返回接收或发送的数据长度。

```
public int getPort()
```

说明：返回发送或接收数据包的远程主机端口。

```
public void setAddress(InetAddress iaddr)
```

说明：设置发送或接收数据包的主机地址。

```
public void setData(byte[] buf)
```

说明：设置数据包内容。

```
public void setLength(int length)
```

说明：设置接收或发送的数据长度。

```
public void setPort(int iport)
```

说明：设置发送或接收数据包的远程主机端口。

9.2.3 DatagramSocket 类

DatagramSocket 类的主要构造方法如下：

```
public DatagramSocket(int port) throws SocketException
public DatagramSocket(int port,InetAddress laddr) throws SocketException
public DatagramSocket(SocketAddress bindaddr) throws SocketException
```

DatagramSocket 类的常用方法如下：

```
public void connect(InetAddress address, int port)
```

说明：建立套接字连接。

```
public void disconnect()
```

说明：断开套接字连接。

```
public InetAddress getInetAddress()
```

说明：返回已连接套接字的地址。

```
public InetAddress getLocalAddress()
```

说明：返回套接字绑定的本地地址。

`public int getLocalPort()`

说明：返回套接字绑定的本地端口。

`public int getPort()`

说明：返回已连接套接字的端口。

`public void receive(DatagramPacket p) throws IOException`

说明：接收数据包。

`public void send(DatagramPacket p) throws IOException`

说明：发送数据包。

9.2.4 创建 UDP 服务器端程序

本节将使用前面介绍的 DatagramSocket 和 DatagramPacket 类创建一个 UDP 服务器端程序。

服务器端接收客户端发出来的空数据包（代表客户端发出请求），由接收的数据包获得客户端的 IP 地址和端口号。然后将服务器端的当前时间以数据包的形式发给客户端，当超过 10 个客户端请求后，服务器端自动关闭。

```java
import java.io.IOException;
import java.net.DatagramPacket;
import java.net.DatagramSocket;
import java.net.InetAddress;
import java.text.SimpleDateFormat;
import java.util.Date;
public class UDPServer {
    private DatagramSocket socket = null;
    private int counter = 1;
    public UDPServer() throws IOException {
        socket = new DatagramSocket(9080);
    }
    public void run() {
        SimpleDateFormat formatter = new SimpleDateFormat("yyyy-MM-dd HH:mm:ss");
        try {
            do {
                byte[] buf = new byte[19];
                DatagramPacket packet = new DatagramPacket(buf,
```

```
            buf.length);
                        socket.receive(packet);
                        String time = formatter.format(new Date());
                        buf = time.getBytes();
                        InetAddress address = packet.getAddress();
                        int port = packet.getPort();
                        packet = new DatagramPacket(buf, buf.length,
address,
   port);
                        socket.send(packet);
                } while (counter < 10);
            } catch (IOException e) {
                e.printStackTrace();
            }
            socket.close();
    }
    public static void main(String[] args) {
        try {
            System.out.println("服务器端已经启动!");
            new UDPServer().run();
            System.out.println("服务器端已经关闭! ");
            System.exit(0);
        } catch (IOException e) {
            e.printStackTrace();
        }
    }
}
```

9.2.5 创建 UDP 客户端程序

本节将使用前面介绍的 DatagramSocket 和 DatagramPacket 类创建一个 UDP 客户端程序。

客户端首先发送请求数据包（空的数据包），然后等待接收服务器端传回来的带有服务器当前时间的数据包，显示服务器端发送时的时间之后关闭。

```
import java.io.IOException;
import java.net.DatagramPacket;
import java.net.DatagramSocket;
import java.net.InetAddress;
import java.net.SocketException;
import java.net.UnknownHostException;
public class UDPClient {
    private DatagramSocket socket = null;
    private String serverIP = "127.0.0.1";
    public UDPClient() throws SocketException {
```

```java
            socket = new DatagramSocket();
        }
        public void setServerIP(String serverIP) {
            this.serverIP = serverIP;
        }
        public void run() {
            try {
                byte[] buf = new byte[19];
                InetAddress address = InetAddress.getByName(serverIP);
                DatagramPacket packet = new DatagramPacket(buf,
buf.length,address, 9080);
                socket.send(packet);
                packet = new DatagramPacket(buf, buf.length);
                socket.receive(packet);
                String received = new String(packet.getData());
                System.out.println("服务器端时间:" + received);
                socket.close();
            } catch (UnknownHostException e) {
                e.printStackTrace();
            } catch (SocketException e) {
                e.printStackTrace();
            } catch (IOException e) {
                e.printStackTrace();
            }
        }
        public static void main(String[] args) {
            try {
                System.out.println("客户端启动,请求获取服务器当前时间的信息...");
                new UDPClient().run();
                System.out.println("客户端已获得服务器当前时间,自动关闭! ");
            } catch (SocketException e) {
                e.printStackTrace();
            }
        }
    }
```

9.3 TCP 协议网络程序

9.3.1 概述

TCP（Tranfer Control Protocol）是一种面向连接的、保证可靠传输的协议。通过 TCP

协议传输，得到的是一个顺序的、无差错的数据流。发送方和接收方成对的两个套接字之间必须建立连接，一旦两个套接字连接起来，它们就可以进行双向数据传输，双方都可以进行发送或接收操作。与 UDP 不同，TCP 对传输数据的大小没有限制。TCP 是一个可靠的协议，其确保接收方完全正确地获取发送方所发送的全部数据。在 java.net 包中提供了两个类：Socket 和 ServerSocket，分别用于表示双向连接的客户端和服务器端。

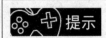 虽然 TCP 是保证可靠传输的协议，但不能表示任何时候都要使用 TCP 协议。应该掌握 TCP 和 UDP 的适用范围从而合理地选择所用协议。

9.3.2 Socket 类

Socket 类的主要构造方法如下：

```
public Socket(InetAddress address,int port) throws IOException
public Socket(InetAddress address,int port,InetAddress localAddr,int localPort) throws IOException
public Socket(String host,int port) throws UnknownHostException,IOException
```

Socket 类的常用方法如下：

```
public InetAddress getInetAddress()
```

说明：返回套接字连接的主机地址。

```
public InetAddress getLocalAddress()
```

说明：返回套接字绑定的本地地址。

```
public InputStream getInputStream() throws IOException
```

说明：获得该套接字的输入流。

```
public int getLocalPort()
```

说明：返回套接字绑定的本地端口。

```
public int getPort()
```

说明：返回套接字连接的远程端口。

```
public OutputStream getOutputStream() throws IOException
```

说明：返回该套接字的输出流。

```
public int getSoTimeout() throws SocketException
```

说明：返回该套接字最长等待时间。

```
public void setSoTimeout(int timeout) throws SocketException
```

说明：设置该套接字最长等待时间。

```
public void shutdownInput() throws IOException
```

说明：关闭输入流。

```
public void shutdownOutput() throws IOexception
```

说明：关闭输出流。

```
public void close() throws IOException
```

说明：关闭套接字。

9.3.3 ServerSocket 类

ServerSocket 类的主要构造方法如下：

```
public ServerSocket(int port) throws IOException
public ServerSocket(int port,int backlog) throws IOException
public ServerSocket(int port,int backlog,InetAddress bindAddr) throws IOException
```

ServerSocket 类的常用方法如下：

```
public Socket accept() throws IOException
```

说明：监听并接受客户端 Socket 连接。

```
public InetAddress getInetAddress()
```

说明：返回服务器套接字的本地地址。

```
public int getLocalPort()
```

说明：返回该套接字监听的端口。

```
public int getSoTimeout() throws SocketException
```

说明：返回该套接字最长等待时间。

```
public void setSoTimeout(int timeout) throws SocketException
```

说明：设置该套接字最长等待时间。

```
public void close() throws IOException
```

说明：关闭套接字。

9.3.4 创建 TCP 服务器端程序

本节将使用前面介绍的 ServerSocket 类创建一个 TCP 服务器端程序。

使用 ServerSocket 监听 9080 端口，等待客户端的连接请求。有客户端建立连接后，接收客户端的信息，然后断开与客户端的连接。当客户端连接次数超过 10 后，关闭服务器端套接字。

```java
import java.io.BufferedReader;
import java.io.IOException;
import java.io.InputStreamReader;
import java.net.ServerSocket;
import java.net.Socket;
public class ServerSocketDemo {
    private ServerSocket ss;
    private Socket socket;
    private BufferedReader in;
    private int counter = 1;
    public void run() {
        try {
            ss = new ServerSocket(9080);
            do {
                socket = ss.accept();
                in = new BufferedReader(new InputStreamReader(socket
                                    .getInputStream()));
                String message = in.readLine();
                System.out.println("接收到客户端" + counter + "发送的消息:" + message);
                in.close();
                socket.close();
                counter++;
            } while (counter < 10);
            ss.close();
        } catch (IOException e) {
            e.printStackTrace();
        }
    }
    public static void main(String[] args) {
        ServerSocketDemo demo = new ServerSocketDemo();
        System.out.println("服务器端已经启动!");
        demo.run();
```

```
            System.out.println(" 服务器端已经关闭！");
            System.exit(0);
        }
    }
}
```

9.3.5 创建 TCP 客户端程序

本节将使用前面介绍的 Socket 类创建一个 TCP 客户端程序。

使用 Socket 连接到地址为 127.0.0.1 的服务器端，端口为 9080。输入一条要发送到服务器端的信息，发送后如果套接字没有关闭，就关闭套接字。

```
import java.io.BufferedReader;
import java.io.IOException;
import java.io.InputStreamReader;
import java.net.ServerSocket;
import java.net.Socket;
public class ServerSocketDemo {
    private ServerSocket ss;
    private Socket socket;
    private BufferedReader in;
    private int counter = 1;
    public void run() {
        try {
            ss = new ServerSocket(9080);
            do {
                socket = ss.accept();
                in = new BufferedReader(new InputStreamReader(socket
                        .getInputStream()));
                String message = in.readLine();
                System.out.println(" 接收到客户端 " + counter + " 发送的消息 :" + message);
                in.close();
                socket.close();
                counter++;
            } while (counter < 10);
            ss.close();
        } catch (IOException e) {
            e.printStackTrace();
        }
    }
    public static void main(String[] args) {
        ServerSocketDemo demo = new ServerSocketDemo();
        System.out.println(" 服务器端已经启动 !");
```

```
        demo.run();
        System.out.println("服务器端已经关闭! ");
        System.exit(0);
    }
}
```

9.4 HTTP 协议网络程序

9.4.1 概述

HTTP 是一个属于应用层的面向对象的协议，适用于分布式超媒体信息系统。HTTP 协议是基于请求/响应范式的。一个客户机与服务器建立连接后，发送一个请求给服务器，请求的格式为统一资源标识符、协议版本号，后面是 MIME 信息（包括请求修饰符、客户机信息和可能的内容）。服务器接收到请求后，给予相应的响应信息，其格式为信息的协议版本号、一个成功或错误的代码，后面是 MIME 信息（包括服务器信息、实体信息和可能的内容）。

9.4.2 URL 类

URL（Uniform Resource Locator）是统一资源定位器的简称，表示 Internet 上某一资源的地址，可以访问 Internet 上的各种网络资源。

URL 的格式如下：

协议名://资源名

其中，协议名为获取资源所使用的传输协议，如 http、ftp、file 等；资源名包括主机名、端口号和文件名。

在 Java 中有一个与 URL 同名的类，用于表示 URL，存在于 java.net 包中。其构造方法如下：

```
public URL(String spec) throws MalformedURLException
public URL(String protocol,String host,int port,String file) throws MalformedURLException
public URL(String protocol,String host,int port,String file,URLStreamHandler handler) throws MalformedURLException
public URL(String protocol,String host,String file) throws MalformedURLException
public URL(URL context,String spec) throws MalformedURLException
public URL(URL context,String spec,URLStreamHandler handler) throws MalformedURLException
```

URL 类中定义了一些方法用解析 URL，如用于获取该 URL 协议名的 getProtocol 方法、用于获取该 URL 主机名的 getHost 方法、用于获取该 URL 端口号的 getPort 方法、用于获取该 URL 文件名的 getFile 方法等。

下面示例演示了如何构造和解析 URL 对象。

```java
import java.net.MalformedURLException;
import java.net.URL;
public class URLDemo {
    public static void main(String[] args) {
        URL url = null;
        try {
            url = new URL("http://www.url.org:8080/demo/info/");
        } catch (MalformedURLException e) {
            e.printStackTrace();
        }
        if (url != null) {
            System.out.println("协议名为" + url.getProtocol());
            System.out.println("主机名为" + url.getHost());
            System.out.println("文件名为" + url.getFile());
            System.out.println("端口号为" + url.getPort());
        }
    }
}
```

这段程序代码的运行结果如下：

```
协议名为 http
主机名为 www.url.org
文件名为 /demo/info/
端口号为 8080
```

除此之外，还可以通过 URL 类中定义的 openStream 方法获得 InputStream 流，从而读取数据。例如：

```java
import java.io.BufferedReader;
import java.io.IOException;
import java.io.InputStreamReader;
import java.net.URL;
public class URLStream {
    public static void main(String[] args) {
        try {
            URL url= new URL("http://www.microsoft.com/");
            BufferedReader in = new BufferedReader(new InputStreamReader(url
```

```
                    .openStream()));
            String inputLine;
            while ((inputLine = in.readLine()) != null)
                System.out.println(inputLine);
            in.close();
        } catch (IOException e) {
            e.printStackTrace();
        }
    }
}
```

执行结果将返回 www.microsoft.com 对应页面 html 代码。

9.4.3 URLConnection 类

在 java.net 包中还有一个 URLConnection 类，其提供了更多的方法。URLConnection 类是一个抽象类，可以通过 URL 类的 openConnection 方法获得。

通过 URLConnection 类不仅可以读取网上数据，也可以输出数据，还提供了检查 HTTP 头的方法。由于方法太多，在此就不一一列举了，感兴趣的读者可以参考 JDK 相关文档。这里只给出一个简单的例子说明其主要方法的使用。

```
import java.io.IOException;
import java.net.URL;
import java.net.URLConnection;
import java.util.Iterator;
import java.util.Map;
import java.util.Set;
public class URLConnectionDemo {
    public static void main(String[] args) {
        try {
            URL url = new URL("http://www.microsoft.com/");
            URLConnection conn = url.openConnection();
            System.out.println("ConnectTimeout:"+ conn.getConnectTimeout());            System.out.println("ReadTimeout:"+ conn.getReadTimeout());
            System.out.println("ContentType"+ conn.getContentType());
            System.out.println("HeaderField Detail:");
            Map map = conn.getHeaderFields();
            Set set = map.entrySet();
            Iterator it = set.iterator();
            while(it.hasNext()){
```

```
                    Map.Entry me = (Map.Entry)it.next();
                    System.out.println(me.getKey()+" "+me.getValue());
                }
            } catch (IOException e) {
                e.printStackTrace();
            }
        }
    }
```

这段程序代码的运行结果如下：

```
ConnectTimeout:0
ReadTimeout:0
ContentTypetext/html; charset=utf-8
HeaderField Detail:
X-Powered-By [ASP.NET]
Content-Length [22613]
X-AspNet-Version [2.0.50727]
Date [Tue, 13 Dec 2010 10:05:17 GMT]
null [HTTP/1.1 200 OK]
Server [Microsoft-IIS/6.0]
Content-Type [text/html; charset=utf-8]
P3P [CP="ALL IND DSP COR ADM CONo CUR CUSo IVAo IVDo PSA PSD TAI TELo OUR SAMo CNT COM INT NAV ONL PHY PRE PUR UNI"]
Cache-Control [private]
```

9.5 综合实例：实现简单的 Web 服务器

在本章的最后，我们将综合运用前面所学的知识给出一个完整的实例。

该实例实现了一个简单的 Web 服务器。实例的源代码如下：

定义 ServerThread 类作为服务器运行的线程主体。

```
import java.io.BufferedReader;
import java.io.DataInputStream;
import java.io.File;
import java.io.FileInputStream;
import java.io.FileNotFoundException;
import java.io.IOException;
import java.io.InputStreamReader;
import java.io.PrintStream;
import java.net.Socket;
```

```java
public class ServerThread extends Thread {
    private static String DEFAULT_FILE_NAME = "index.htm";
    private Socket client;
    public ServerThread(Socket clinet) {
        this.client = clinet;
    }
    public void run() {
        try {
            PrintStream outstream = new PrintStream(client.getOutputStream());
            DataInputStream instream = new DataInputStream(client
                    .getInputStream());
            String inline = new BufferedReader(new InputStreamReader(instream)).readLine();
            System.out.println("Received:" + inline);
            if (isGetRequest(inline)) {
                String filename = getReqFileName(inline);
                File file = new File(filename);
                if (file.exists()) {
                    System.out.println(filename + " requested.");
                    outstream.println("HTTP/1.0 200 OK");
                    outstream.println("MIME_version:1.0");
                    outstream.println("Content_Type:text/html");
                    int len = (int) file.length();
                    outstream.println("Content_Length:" + len);
                    outstream.println("");
                    response(outstream, file);
                    outstream.flush();
                } else {
                    String notfound = "<html><head><title>Not Found</title></head><body><h1>Error 404-file not found</h1></body></html>";
                    outstream.println("HTTP/1.0 404 no found");
                    outstream.println("Content_Type:text/html");
                    outstream     .println("Content_Length:" + notfound.length() + 2);
                    outstream.println("");
                    outstream.println(notfound);
                    outstream.flush();
                }
```

```java
                }
                long m1 = 1;
                while (m1 < 11100000) {
                    m1++;
                }
                client.close();
            } catch (IOException e) {
                e.printStackTrace();
            }
        }
    }
    private boolean isGetRequest(String s) {
        if (s.length() > 0 && s.substring(0, 3).
equalsIgnoreCase("GET")) {
            return true;
        }
        return false;
    }
    private String getReqFileName(String s) {
        String fileName = s.substring(s.indexOf(' ') + 1);
        fileName = fileName.substring(0, fileName.indexOf(' '));
        try {
            if (fileName.charAt(0) == '/')
                fileName = fileName.substring(1);
        } catch (StringIndexOutOfBoundsException e) {
            e.printStackTrace();
        }
        if (fileName.length() == 0)
            fileName = DEFAULT_FILE_NAME;
        return fileName;
    }
    private void response(PrintStream outs, File file) {
        try {
            DataInputStream in = new DataInputStream(new FileInputStream(file));
            int len = (int) file.length();
            byte buf[] = new byte[len];
            in.readFully(buf);
            outs.write(buf, 0, len);
            outs.flush();
            in.close();
        } catch (FileNotFoundException e) {
            e.printStackTrace();
```

```java
        } catch (IOException e) {
            e.printStackTrace();
        }
    }
    public static String getDEFAULT_FILE_NAME() {
        return DEFAULT_FILE_NAME;
    }
    public static void setDEFAULT_FILE_NAME(String default_file_name) {
        DEFAULT_FILE_NAME = default_file_name;
    }
}
```

之后定义 SimpleWebServer 类。

```java
import java.io.IOException;
import java.net.ServerSocket;
public class SimpleWebServer {
    private static int Counter = 1;
    private static int port = 8090;
    public static void run() throws IOException {
        ServerSocket server = new ServerSocket(port);
        System.out.println("<--SimpleWebServer 正在监听 " +
server.getLocalPort()+ " 端口-->");
        while (true) {
            new ServerThread(server.accept()).start();
            Counter++;
        }
    }
    public static int getPort() {
        return port;
    }
    public static void setPort(int port) {
        SimpleWebServer.port = port;
    }
    public static int getCounter() {
        return Counter;
    }
    public static void main(String[] args) {
        try {
            SimpleWebServer.run();
        } catch (IOException e) {
            System.out.println("<--SimpleWebServer 出现异常-->");
            e.printStackTrace();
```

```
        }
    }
}
```

9.6 要点总结

本章首先介绍了 OSI 和 TCP/IP 网络模型及套接字的概念，然后分别介绍了基于 UDP、TCP 和 HTTP 协议下基本的网络编程方法。

9.7 练习题

1．填空题

（1）OSI 参考模型由 7 层组成，它们分别是物理层、数据链路层、网络层、传输层、_____、_____ 和 _____。

（2）与 OSI 参考模型不同，TCP/IP 参考模型只有 4 层，从下向上依次是网络接口层、_____、传输层和 _____。

（3）套接字之间的连接过程可以分为三个步骤：_____、_____、_____。

（4）UDP 是一种 _____ 的协议，而 TCP 是一种 _____ 的、保证可靠传输的协议。

（5）HTTP 是一个属于应用层的 _____ 协议，适用于分布式超媒体信息系统。

2．选择题

（1）下列哪个类用于在程序之间建立传送数据包的通信连接 _____。

    A. DatagramPacket　　B. DatagramSocket　　C. Socket　　D. ServerSocket

（2）下列哪个类用于表示一个数据包 _____。

    A. DatagramPacket　　B. DatagramSocket　　C. Socket　　D. ServerSocket

（3）下列哪个类用于表示 TCP 客户端套接字 _____。

    A. DatagramPacket　　B. DatagramSocket　　C. Socket　　D. ServerSocket

（4）下列哪个类用于表示 TCP 服务端套接字 _____。

    A. DatagramPacket　　B. DatagramSocket　　C. Socket　　D. ServerSocket

（5）下列哪个类提供了检查 HTTP 头的方法 _____。

    A. URL　　B. URLConnection　　C. Socket　　D. ServerSocket

3. 问答题

（1）简述 UDP 和 TCP 协议的异同。

（2）简述基于 TCP 协议下的客户 / 服务器端套接字的工作原理。

9.8 编程练习

请模仿本章中的示例，完成一个基于 UDP 或 TCP 协议的网络通信程序。

第 10 章
Java 图形用户界面

Java 为图形用户界面（Grahi User Interface，GUI）提供的对象都保存在java Awt 和 javx.Swing 两个包中，包内有相关的组件。GUI 能够用图形的方式显示计算机操作界面，这样更加方便、直观。本章主要介绍 GUI 的相关组件。

10.1 AWT 与 Swing 简介

Swing 是 Java 为图形界面应用开发提供的一组工具包，是 Java 基础类的一部分。Swing 包含了构建图形界面（GUI）的各种组件，如窗口、标签、按钮、文本框等。Swing 提供了许多比 AWT 更好的屏幕显示元素，使用纯 Java 实现，能够更好地兼容跨平台运行。

10.1.1 AWT 简介

AWT 是抽象窗口工具包（Abstract Window Toolkit）的英文缩写。抽象窗口工具包为开发者提供了建立图形用户界面的工具集。其主要功能如下：

（1）用户界面组件。

（2）事件处理模型。

（3）图形和图像工具。

（4）布局管理器。

（5）数据传送类。

AWT 主要涉及 java.awt 包，这个包也是 java 众多包中最大的一个。java.awt 包中提供了图形用户界面设计所使用的类和接口。提供的工具类主要包括以下三种：

（1）组件（Component）：用户在图形界面中用户经常看到的按钮、标签、菜单等就是组件。

（2）容器（Container）：所有的 AWT 组件都放在容器中，并可以设置其位置、大小等。

（3）布局管理器（LayoutManager）：可以使容器中的组件按照指定位置进行摆放。

10.1.2 Swing 简介

在 Java 刚诞生的时候，AWT 是 Java 唯一用于开发图形界面的基本编程库，但随着 Java 的广泛应用，AWT 逐渐暴露了其自身的不足，如缺少剪贴板、打印支持、键盘导航等，以及基于本地对等组件的同位体体系结构更是容易在不同操作系统下出现问题。AWT 逐渐不能满足图形界面开发的需求，这也正是促进 Swing 产生的根本原因。于是在 1996 年，sun 与 Netscape 合作创建了 Swing 库。

Swing 的组件几乎都是轻量级的，与 AWT 组件不同的是，这些组件没有本地对等组件是由纯 Java 实现的，因此它们不依赖于操作系统。与 AWT 的重量级组件相比，Swing 组件被称作轻量级组件。重量级组件是在本地的不透明窗体中绘制，而轻量级组件是在重量级组件的窗口中绘制。由于抛弃了基于本地对等组件的同位体体系结构，因此 Swing 不但在不同的平台上表现一致，而且还提供了本地组件不支持的特性。

然而，Swing 的出现并不代表 AWT 的设计就是失败的。需要明确的是，AWT 是 Swing 的基础，AWT 最初的设计也只是定位于小应用程序的简单用户界面。

使用 Swing 开发图形界面，所有的组件、容器和布局管理器都在 javax.swing 包中。

10.1.3 容器简介

组件不能独立显示出来，必须将组件放在一定的容器中。容器不仅可以容纳组件，也可以容纳容器。实际上容器是一种特殊的组件，具备组件的所有性质，但是其主要功能是容纳其他组件和容器。

java.awt.Container 类是 java.awt.Component 的子类，一个容器可以容纳多个组件，并使它们成为一个整体。所有的容器都可以通过 add() 方法向容器中添加组件。有三种类型的容器，即 Window、Panel、ScrollPane，常用的有 Panel、Frame 等。

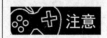 容器也是一种组件，具备一般组件的所有性质。

10.2 创建窗体

在开发 Java 应用程序时，通常利用 JFrame 类创建窗体。利用 JFrame 类创建的窗体分别包含一个标题、最小化按钮、最大化按钮和关闭按钮。JFrame 类提供了一系列用于设置窗体的方法，常用的操作方法如表 10-1 所示。

表10-1 JFrame类中的常用操作方法

序号	方 法	描 述
1	public JFrame() throws HeadlessException	构造一个不可见的窗体对象
2	public JFrame(String title) throws HeadlessException	构造一个带标题的窗体对象
3	public void setSize(int width, int height)	设置窗体大小
4	public void setSize(Dimension d)	通过Dimension设置窗体大小
5	public void setBackground(Color c)	设置窗体的背景颜色
6	public void setLocation(int x, int y)	设置组件类的显示位置
7	public void setBounds(int x,int y,int width, int height)	设置组件类的显示边界
8	public void setLocation(Point p)	通过Point设置组件的显示位置
9	public void setVisible(boolean b)	显示或隐藏组件
10	public Component add(Component comp)	向容器中增加组件
11	public void setLayout(LayoutManager manager)	设置布局管理器，null不设置
12	public Container getContentPane()	返回此窗体的容器对象

下面使用以上方法创建一个新窗体。

```
import java.awt.Color;
import javax.swing.JFrame;
public class JFrameDemo01 {
    public static void main(String[] args) {
        JFrame f = new JFrame("第一个Swing窗体");    // 实例化窗体对象
        f.setSize(230, 160);                        // 设置窗体大小
        f.setBackground(Color.WHITE);               // 设置窗体的背景颜色
        f.setLocation(300, 200);                    // 设置窗体的显示位置
        f.setVisible(true);                         // 让组件显示
        // 设置关闭按钮的动作作为关闭窗体
        f.setDefaultCloseOperation(JFrame.EXIT_ON_CLOSE);
    }
}
```

程序运行结果如图 10-1 所示。

图 10-1 程序运行结果

 利用 JFrame 类创建窗体时，必须在最后设置为可见（默认情况下窗体不可见）。为了让窗体的关闭按钮可用，必须设置关闭按钮为可用，否则需要使用 Ctrl+C 组合键退出程序。

也可以使用继承 JFrame 类的方法创建窗体。例如：

```java
import java.awt.Container;
import javax.swing.JFrame;
public class JFrameDemo02 extends JFrame {
    public JFrameDemo02(){
        createUserInterface();
    }
    public void createUserInterface(){
        Container contentPane = getContentPane();   // 取得窗体的容器
        contentPane.setLayout( null );              // 不使用任何布局管理器
        setTitle("第一个 Swing 窗体");                // 设置窗体的标题
        setBounds(300, 200, 230, 160);              // 设置窗体显示位置和大小
        setVisible(true);                           // 让组件显示
    }
    public static void main(String[] args) {
        JFrameDemo02 frame = new JFrameDemo02();
        frame.setDefaultCloseOperation(JFrame.EXIT_ON_CLOSE);
    }
}
```

程序运行结果与上例相同。

10.3 标签组件：JLabel

JLabel 组件表示的是一个标签，本身是用来显示信息的，一般情况下不能更改其显示内容。创建完的 JLabel 对象可以通过 Container 类中的 add() 方法加入到容器中。JLabel 类中的常用方法和常量如表 10-2 所示。

表10-2 JLabel类中的常用方法和常量

序号	方法及常量	类型	描述
1	public static final int LEFT	常量	标签文本左对齐
2	public static final int RIGHT	常量	标签文本右对齐
3	public static final int CENTER	常量	标签文本居中对齐
4	public JLabel()	构造	创建一个JLabel对象
5	public JLabel(String text)	构造	创建一个指定文本内容的JLabel对象，默认左对齐

（续表）

序号	方法及常量	类型	描述
6	public JLabel(String text, int Alignment)	构造	创建一个指定文本内容和对齐方式的JLabel对象
7	public JLabel(String text,Icon icon, int horizontalAlignment)	构造	创建具有指定文本、图像和水平对齐方式的JLabel对象
8	public void setText(String text)	普通	设置标签的文本
9	public String getText()	普通	取得标签的文本
10	public void setIcon(Icon icon)	普通	设置指定的图像
11	public void setAlignment(int alignment)	普通	设置标签的对齐方式

下面是使用标签的示例：

```
import java.awt.Color;
import java.awt.Dimension;
import java.awt.Point;
import javax.swing.JFrame;
import javax.swing.JLabel;
public class JLabelDemo01 {
    public static void main(String[] args) {
        JFrame f = new JFrame("JLabel示例窗体");// 实例化窗体对象
        // 实例化对象，居中对齐
JLabel label = new JLabel("周口师范学院",JLabel.CENTER);
        f.add(label);                        // 向容器中加入组件
        Dimension dim = new Dimension();     // 实例化对象
        dim.setSize(230, 160);               // 设置大小
        f.setSize(dim);                      // 设置组件大小
        f.setBackground(Color.WHITE);        // 设置窗体的背景颜色
        Point point = new Point(300, 200);   // 设置显示的坐标点
        f.setLocation(point);                // 设置窗体的显示位置
        f.setVisible(true);                  // 让组件显示
        // 设置关闭按钮关闭窗体
f.setDefaultCloseOperation(JFrame.EXIT_ON_CLOSE);
    }
}
```

程序运行结果如图10-2所示。

图 10-2 程序运行结果

上面示例的标签内容只是使用了默认的字体及颜色显示，如果现在需要更改使用的字体，就可以直接使用 java.awt.Font 类来表示字体。常用的方法如下：

```
    protected Font(Font font)              // 设置字体
    public Font(String name,int style,int size)
```

根据指定名称、样式和磅值大小创建一个新 Font：

```
import java.awt.Color;
import java.awt.Dimension;
import java.awt.Font;
import java.awt.Label;
import java.awt.Point;
import javax.swing.JFrame;
import javax.swing.JLabel;
public class JLabelDemo02 {
    public static void main(String[] args) {
        JFrame f = new JFrame("JLabel示例窗体 ");    // 实例化窗体对象
        JLabel lab = new JLabel(" 周口师范学院 ");      // 实例化对象
        lab.setBounds(0, 0, 100, 100);              // 设置标签的显示位置和大小
        // 设置文本的字体和大小
lab.setFont(new Font("Serief",Font.BOLD+Font.ITALIC,23));
        lab.setHorizontalAlignment(Label.LEFT);// 设置水平对齐方式
        lab.setForeground(Color.RED);               // 设置标签的文字颜色
        f.add(lab);                                 // 向容器中加入组件
        Dimension dim = new Dimension();            // 实例化对象
        dim.setSize(230, 160);                      // 设置大小
        f.setSize(dim);                             // 设置组件大小
        f.setBackground(Color.WHITE);               // 设置窗体的背景颜色
        Point point = new Point(300, 200);          // 设置显示的坐标点
        f.setLocation(point);                       // 设置窗体的显示位置
        f.setVisible(true);                         // 让组件显示
        // 设置关闭按钮关闭窗体
f.setDefaultCloseOperation(JFrame.EXIT_ON_CLOSE);
    }
}
```

程序运行结果如图 10-3 所示。

图 10-3 程序运行结果

在 JLabel 类中可以设置图片，直接使用 Icon 接口及 ImageIcon 子类即可，在 ImageIcon

类中使用构造方法（表 10-3）将图像的数据以 byte 数组的形式设置上去。

表10-3 ImageIcon类的构造方法

序号	构造方法	描述
1	public ImageIcon(String filename)	通过文件名创建对象
2	public ImageIcon(String filename, String description)	设置图片路径及图片描述
3	public ImageIcon(byte[] imageData)	将保存图片信息的byte数组设置到ImageIcon中

下面的示例演示了从文件中读取图片并在标签上显示的方法。

```java
import java.awt.Color;
import java.io.File;
import javax.swing.Icon;
import javax.swing.ImageIcon;
import javax.swing.JFrame;
import javax.swing.JLabel;
public class JLabelDemo03 {
    public static void main(String[] args) {
        JFrame f = new JFrame("JLabel 图像示例窗体 ");// 实例化窗体对象
        //Icon icon = new ImageIcon("China.png");
        String picPath = "d:" + File.separator +"China.png";
        Icon icon = new ImageIcon(picPath);       // 实例化 Icon 对象
        // 实例化标签对象
        JLabel label = new JLabel(" 国旗 ",icon,JLabel.CENTER);
        label.setBackground(Color.YELLOW);       // 设置背景颜色
        label.setForeground(Color.RED);          // 设置标签的文字颜色
        f.add(label);                            // 向容器中加入组件
        f.setSize(260, 160);                     // 设置大小
        f.setBackground(Color.WHITE);            // 设置窗体的背景颜色
        f.setLocation(300, 200);                 // 设置窗体的显示位置
        f.setVisible(true);                      // 让组件显示
        // 设置关闭按钮关闭窗体
        f.setDefaultCloseOperation(JFrame.EXIT_ON_CLOSE);
    }
}
```

程序运行结果如图 10-4 所示。

图 10-4 程序运行结果

> **提示** 在程序中直接通过文件名实例化 ImageIcon 对象时（如程序中注释的语句），文件的路径应放在 eclipse 项目的根目录下，否则会找不到。另外，如果图像来源于一个不确定输入流（如从数据库中读取 BLOB 字段），就需要通过 InputStream 完成操作。

10.4 按钮组件：JButton

JButton 组件表示一个普通按钮，使用此类可以直接在窗体中增加一个按钮。JButton 类中常用的方法如表 10-4 所示。

表10-4 JButton类中的常用方法

序号	方 法	描 述
1	public JButton()	构造一个JButton对象
2	public JButton(String text)	创建一个带文本的按钮
3	public JButton(Icon icon)	创建一个带图标的按钮
4	public JButton(String text,Icon icon)	创建带初始文本和图标的按钮
5	public void setMnemonic(int mnemonic)	设置按钮的快捷键
6	public void setText(String text)	设置JButton的显示内容

JButton 组件只是在按下和释放两个状态之间进行切换，可以通过捕获按下并释放的动作执行一些操作，从而完成和用户的交互。下面的示例演示了创建一个按钮的方法。

```java
import javax.swing.Icon;
import javax.swing.ImageIcon;
import javax.swing.JButton;
import javax.swing.JFrame;
public class JButtonDemo {
    public static void main(String[] args) {
        JFrame f = new JFrame("JButton 示例窗体 ");  // 实例化窗体对象
        f.setLayout(null);                          // 不使用布局管理器
        JButton b1 = new JButton();                 // 定义按钮对象
        b1.setText(" 按我 ");                        // 设置按钮的显示文本
        b1.setBounds(0, 30, 100, 30);               // 设置按钮的位置及大小
        Icon icon = new ImageIcon("China.png");     // 实例化 Icon 对象
        JButton b2 = new JButton(icon);             // 定义按钮对象
        b2.setBounds(110, 10, 130, 100);            // 设置按钮的位置及大小
        f.add(b1);                                  // 向容器中加入组件
        f.add(b2);                                  // 向容器中加入组件
        f.setSize(260, 160);                        // 设置大小
        f.setLocation(300, 200);                    // 设置窗体的显示位置
        f.setVisible(true);                         // 让组件显示
```

```
        // 设置关闭按钮关闭窗体
        f.setDefaultCloseOperation(JFrame.EXIT_ON_CLOSE);
    }
}
```

程序运行结果如图 10-5 所示。

图 10-5 程序运行结果

从结果可以发现，为按钮设置一张显示图片的方法与 JLabel 类似。

10.5 JPanel 容器

JPanel 容器是一种常用的容器之一，可以使用它完成各种复杂的界面显示。在 JPanel 容器中可以加入任意组件，之后直接将 JPanel 容器加入到 JFrame 容器中即可显示。下面的示例演示了 JPanel 容器的基本使用。

```java
import javax.swing.JButton;
import javax.swing.JFrame;
import javax.swing.JLabel;
import javax.swing.JPanel;
public class JPanelDemo {
    public static void main(String[] args) {
        JFrame f = new JFrame("JPanel 示例");      // 实例化窗体对象
        JPanel p = new JPanel();                    // 实例化 JPanel 对象
        p.add(new JLabel("标签-A"));
        p.add(new JLabel("标签-B"));
        p.add(new JLabel("标签-C"));
        p.add(new JButton("按钮-X"));
        p.add(new JButton("按钮-Y"));
        p.add(new JButton("按钮-Z"));
        f.pack();                                   // 根据组件自动调整窗体大小
        f.setSize(130, 100);                        // 设置大小
        f.setLocation(300, 200);                    // 设置窗体的显示位置
        f.setVisible(true);                         // 让组件显示
        // 设置关闭按钮关闭窗体
        f.setDefaultCloseOperation(JFrame.EXIT_ON_CLOSE);
```

        }
    }

从结果可以发现,所有的组件采用顺序形式加入到了 JPanel 容器中,最后将 JPanel 容器加入到 JFrame 中。JPanel 容器结合布局管理器可以更加方便地管理组件。

10.6 布局管理器

每个容器都有自己的布局管理器,用于对容器内的组件进行定位、设置大小、排列顺序等。因为使用布局管理器是为了让生成的图形用户界面具有良好的平台无关性,所以建议使用布局管理器管理容器内组件的布局和大小。不同的布局管理器使用不同的算法和策略,容器可以通过选择不同的布局管理器来决定布局。布局管理器主要包括 FlowLayout、BorderLayout、GridLayout、CardLayout。前面使用的 setBound(int x,int y,int width,int height) 是通过设置绝对坐标的方式完成的,称为绝对定位。

 布局管理器是实现图形用户界面平台无关性的关键。

10.6.1 FlowLayout

FlowLayout 属于流式布局管理器,它的布局方式是先在一行排列组件,当该行没有足够的空间时,再回行显示。下面的示例演示了 FlowLayout 的设置方法。

```
import java.awt.FlowLayout;
import javax.swing.JButton;
import javax.swing.JFrame;
public class FlowLayoutDemo {
    public static void main(String[] args) {
        JFrame f = new JFrame("FlowLayout 示例 ");        // 实例化窗体对象
    // 设置窗体的布局管理器为 FlowLayout,所有组件居中对齐,水平和垂直间距为 3
        f.setLayout(new FlowLayout(FlowLayout.CENTER,3,3));
        JButton b = null;
        for (int i = 0; i < 8; i++) {
            b = new JButton(" 按钮 -"+ i);
            f.add(b);
        }
        f.setSize(230, 130);// 设置大小
        f.setLocation(300, 200);// 设置窗体的显示位置
        f.setVisible(true);// 让组件显示
    // 设置关闭按钮关闭窗体
        f.setDefaultCloseOperation(JFrame.EXIT_ON_CLOSE);
    }
}
```

程序运行结果如图 10-6 所示。

图 10-6 程序运行结果

从程序运行结果中可以发现，所有的组件按照顺序依次向下排列，每个组件之间的间距是 3，居中对齐。

10.6.2 BorderLayout

BorderLayout 将一个窗体的版面划分为东、西、南、北、中 5 个区域，可以直接将需要的组件放到这 5 个区域中。BorderLayout 是 JFrame 窗体的默认布局管理器，其布局方式如图 10-7 所示。

图 10-7 BorderLayout 布局方式

如果组件容器采用了边界布局管理器，那么在将组件添加到容器时，需要设置组件的显示位置，通过 add() 方法添加。下面的示例演示了 BorderLayout 布局管理器的使用方法。

```java
import java.awt.BorderLayout;
import javax.swing.JButton;
import javax.swing.JFrame;
public class BorderLayoutDemo {
    public static void main(String[] args) {
        JFrame f = new JFrame("BorderLayout 示例 ");    // 实例化窗体对象
        // 设置窗体的布局管理器为BorderLayout，所有组件水平和垂直间距为3
        f.setLayout(new BorderLayout(3,3));
        f.add(new JButton(" 东 (EAST)"), BorderLayout.EAST);
        f.add(new JButton(" 西 (WEST)"), BorderLayout.WEST);
        f.add(new JButton(" 南 (SOUTH)"), BorderLayout.SOUTH);
        f.add(new JButton(" 北 (NORTH)"), BorderLayout.NORTH);
        f.add(new JButton(" 中 (CENTER)"), BorderLayout.CENTER);
        f.pack();                          // 根据组件自动调整窗体大小
        f.setSize(300, 160);               // 设置大小
```

```
            f.setLocation(300, 200);              // 设置窗体的显示位置
            f.setVisible(true);                   // 让组件显示
            // 设置关闭按钮关闭窗体
            f.setDefaultCloseOperation(JFrame.EXIT_ON_CLOSE);
    }
}
```

程序运行结果如图 10-8 所示。

图 10-8 程序运行结果

10.6.3 GridLayout

GridLayout 为网格布局管理器,它的布局方式是将容器按照用户的设置平均划分若干网,以表格的形式进行管理。在使用该布局管理器时,必须设置显示的行数和列数。

下面的示例是通过实现计算器面板来演示 GridLayout 布局管理器的使用方法。

```
import java.awt.GridLayout;
import javax.swing.JButton;
import javax.swing.JFrame;
public class GridLayoutDemo {
    public static void main(String[] args) {
        JFrame f = new JFrame("计算器面板示例");  // 实例化窗体对象
        // 设置窗体的布局管理器为 GridLayout,按 4X4 进行排列,
// 所有组件水平和垂直间距为 3
        f.setLayout(new GridLayout(4,4,3,3));
        String[][] names = {{"1","2","3","+"},{"4","5","6","-"},
{"7","8","9","*"},{".","0","=","/"}};
        JButton[][] b = new JButton[4][4];
        for (int row = 0; row < names.length; row++) {
            for (int col = 0; col < names.length; col++) {
                // 创建按钮对象
b[row][col] = new JButton(names[row][col]);
                f.add(b[row][col]);
            }
        }
        f.pack();// 根据组件自动调整窗体大小
```

```
            f.setSize(300, 160);// 设置大小
            f.setLocation(300, 200);// 设置窗体的显示位置
            f.setVisible(true);// 让组件显示
            // 设置关闭按钮关闭窗体
            f.setDefaultCloseOperation(JFrame.EXIT_ON_CLOSE);
        }
    }
```

程序运行结果如图 10-9 所示。

图 10-9 程序运行结果

10.6.4 CardLayout

CardLayout 是将一组组件彼此重叠地进行布局，就像一张张卡片一样。由于每次只会展现一个界面，因此 CardLayout 布局管理器还需要有用于翻转的方法。例如：

```
import java.awt.CardLayout;
import java.awt.Container;
import javax.swing.JFrame;
import javax.swing.JLabel;
public class CardLayoutDemo {
    public static void main(String[] args) {
        JFrame f = new JFrame("CardLayout 示例 ");         // 实例化窗体对象
        Container c = f.getContentPane();                  // 取得窗体容器
        CardLayout card = new CardLayout();// 定义布局管理器
        f.setLayout(card);// 设置布局管理器
        c.add(new JLabel("First",JLabel.CENTER),"first");
        c.add(new JLabel("Second",JLabel.CENTER),"second");
        c.add(new JLabel("Third",JLabel.CENTER),"third");
        f.pack(); // 根据组件自动调整窗体大小
        f.setSize(130, 100);// 设置大小
        f.setLocation(300, 200);// 设置窗体的显示位置
        f.setVisible(true);// 让组件显示
        card.show(c, "second");// 显示第 2 张卡片
        for (int i = 0; i < 3; i++) {
            try {
                Thread.sleep(3000); // 加入显示延迟
            } catch (InterruptedException e) {
```

```
                e.getStackTrace();
            }
            card.next(c);                    // 从容器中取出组件
        }
        // 设置关闭按钮关闭窗体
        f.setDefaultCloseOperation(JFrame.EXIT_ON_CLOSE);
    }
}
```

程序运行结果如图 10-10 所示。

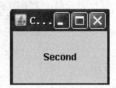

图 10-10 程序运行结果

运行时，首先会显示第 2 张卡片，之后循环显示每一张卡片。

10.7 文本组件：JTextComponent

在 Swing 中提供了三类文本输入组件：

- 单行文本框：JTextField；
- 密码文本框：JPasswordField；
- 多行文本框：JTextArea。

在程序开发中，JTextComponent 组件的常用方法如表 10-5 所示。

表10-5 JTextComponent组件的常用方法

序号	方法	描述
1	public String getText()	返回文本框的所有内容
2	public String getSelectedText()	返回文本框中选定的内容
3	public int getSelectionStart()	返回选定文本的起始位置
4	public int getSelectionEnd()	返回选定文本的结束位置
5	public void selectAll()	选择此文本框的所有内容
6	public void setText(String t)	设置此文本框的内容
7	public void select(int selectionStart, int selectionEnd)	将指定范围内的内容选定
8	public void setEditable(boolean b)	设置此文本框是否可编辑

10.7.1 单行文本框：JTextField

JTextField 组件实现一个文本框，用来接收用户输入的单行文本信息，可以设置默认文本、文本长度、文本字体、文本格式等。JTextField 组件的常用方法如表 10-6 所示。

表10-6 JTextField组件的常用方法

序号	方法	描述
1	public JTextField()	构造默认的文本框
2	public JTextField(String text)	构造指定文本内容长度文本框
3	public JTextField(int columns)	设置文本框内容的长度
4	public JTextField(String text,int columns)	构造指定文本内容并设置长度
5	public void setFont(Font f)	设置文本框文本的字体
6	public void setHorizontalAlignment(int alignment)	设置文本的水平对齐方式

下面的示例演示了文本框的使用方法。

```java
import java.awt.Font;
import java.awt.GridLayout;
import javax.swing.JFrame;
import javax.swing.JLabel;
import javax.swing.JTextField;
public class JTextFieldDemo {
    public static void main(String[] args) {
        JFrame f = new JFrame("JTextField示例 ");        // 实例化窗体对象
        JLabel lb1 = new JLabel(" 姓名：");               // 创建标签对象
        // 定义文本框，指定内容和长度
        JTextField text1 = new JTextField(" 陈占伟 ",30);
        text1.setFont(new Font("",Font.BOLD,12));        // 设置文本的字体
        JLabel lb2 = new JLabel(" 部门：");               // 创建标签对象
        // 定义文本框并指定内容
        JTextField text2 = new JTextField(" 计算机科学系 ");
        // 设置文本框内容的水平对齐方式
        text2.setHorizontalAlignment(JTextField.CENTER);
        text2.setEditable(false);                        // 设置文本框不可编辑
        f.setLayout(new GridLayout(2, 2));               // 设置布局管理器
        f.add(lb1);                                      // 向容器中添加组件
        f.add(text1);                                    // 向容器中添加组件
        f.add(lb2);                                      // 向容器中添加组件
        f.add(text2);                                    // 向容器中添加组件
        f.pack();                                        // 根据组件自动调整窗体大小
        f.setSize(300, 100);                             // 设置大小
        f.setLocation(300, 200);                         // 设置窗体的显示位置
        f.setVisible(true);                              // 让组件显示
        // 设置关闭按钮关闭窗体
        f.setDefaultCloseOperation(JFrame.EXIT_ON_CLOSE);
    }
}
```

程序运行结果如图 10-11 所示。

图 10-11 程序运行结果

10.7.2 密码文本框：JPasswordField

JPasswordField 组件实现一个密码框，用来接收用户输入的单行文本信息，但是在密码框中并不显示输入的真实信息，而是通过显示一个指定的回显字符作为占位符，常见的默认回显字符为 "*"。JPasswordField 组件的常用方法如表 10-7 所示。

表10-7 JPasswordField组件的常用方法

序号	方法	描述
1	public JPasswordField()	构造默认的JPasswordField对象
2	public JPasswordField(String text)	构造指定内容的JPasswordField对象
3	public void setEchoChar(char c)	设置回显字符，默认为 "*"
4	public char getEchoChar()	获得回显字符，返回值为char
5	public char[] getPassword()	获得此文本框的所有内容

下面的示例演示了设置回显字符的方法。

```
import javax.swing.JFrame;
import javax.swing.JLabel;
import javax.swing.JPasswordField;
public class JPasswordFieldDemo {
    public static void main(String[] args) {
        JFrame f = new JFrame("JPasswordField示例");// 实例化窗体对象
        JPasswordField pw1 = new JPasswordField(); // 定义密码文本框
        JPasswordField pw2 = new JPasswordField();// 定义密码文本框
        pw2.setEchoChar('#');                    // 设置回显字符 "#"
        JLabel lb1 = new JLabel("默认回显: ");    // 创建标签对象
        JLabel lb2 = new JLabel("回显设置 #: ");  // 创建标签对象
        lb1.setBounds(10, 10, 100, 20);          // 设置组件位置及大小
        lb2.setBounds(10, 40, 100, 20);          // 设置组件位置及大小
        pw1.setBounds(110, 10, 80, 20);          // 设置组件位置及大小
        pw2.setBounds(110, 40, 50, 20);          // 设置组件位置及大小
        f.setLayout(null);                       // 使用绝对定位
        f.add(lb1);                              // 向容器中添加组件
        f.add(lb2);                              // 向容器中添加组件
        f.add(pw1);                              // 向容器中添加组件
        f.add(pw2);                              // 向容器中添加组件
        f.pack();                                // 根据组件自动调整窗体大小
        f.setSize(300, 100);                     // 设置大小
        f.setLocation(300, 200);                 // 设置窗体的显示位置
```

```
            f.setVisible(true);                    // 让组件显示
            // 设置关闭按钮关闭窗体
            f.setDefaultCloseOperation(JFrame.EXIT_ON_CLOSE);
    }
}
```

程序运行结果如图 10-12 所示。

图 10-12 程序运行结果

10.7.3 多行文本框：JTextArea

JTextArea 组件实现多行文本的输入，也称为文本域。在创建文本域时，可以设置是否自动换行，默认为 false。如果一个文本域太大，就会使用滚动条显示，此时需要将文本域设置在带滚动条的面板中，可以使用 JScrollPane。JScrollPane 组件可以设置水平滚动条和垂直滚动条。水平滚动条根据需要来显示，而垂直滚动条将始终显示。JTextArea 组件的常用方法如表 10-8 所示。

表10-8 JTextArea组件的常用方法

序号	方 法	描 述
1	public JTextArea()	构造文本域，行数和列数为0
2	public JTextArea(int rows,int columns)	构造文本域，指定行数和列数
3	public JTextArea(String text,int rows,int columns)	指定文本域的内容、行数和列数
4	public void append(String str)	在文本域中追加内容
5	public void insert(String str, int pos)	在指定位置插入文本
6	public void setLineWrap(boolean wrap)	设置换行策略

下面的示例演示了 JTextArea 和 JScrollPane 的使用方法。

```
import java.awt.GridLayout;
import javax.swing.JFrame;
import javax.swing.JLabel;
import javax.swing.JScrollPane;
import javax.swing.JTextArea;
public class JTextAreaDemo {
    public static void main(String[] args) {
        JFrame f = new JFrame("JTextArea示例");// 实例化窗体对象
        JTextArea tArea = new JTextArea(3,20);          // 构造文本域
        tArea.setLineWrap(true);                 // 如果内容过多，就自动换行
```

```
            // 在文本域上加入滚动条，水平和垂直滚动条始终出现
            JScrollPane scroll = new JScrollPane(tArea,
JScrollPane.VERTICAL_SCROLLBAR_ALWAYS,
JScrollPane.HORIZONTAL_SCROLLBAR_ALWAYS);
            JLabel lb = new JLabel("多行文本域：");    // 定义标签
            f.setLayout(new GridLayout(2, 1));        // 设置布局管理器
            f.add(lb);                                // 向容器中添加组件
            f.add(scroll);                            // 向容器中添加组件
            f.pack();                                 // 根据组件自动调整窗体大小
            f.setSize(300, 100);                      // 设置大小
            f.setLocation(300, 200);                  // 设置窗体的显示位置
            f.setVisible(true);                       // 让组件显示
            // 设置关闭按钮关闭窗体
            f.setDefaultCloseOperation(JFrame.EXIT_ON_CLOSE);
    }
}
```

程序运行结果如图 10-13 所示。

从程序的运行结果中可以清楚地发现，如果一个文本域中的内容过多，就自动进行换行显示。

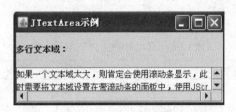

图 10-13 程序运行结果

10.8 事件处理

一个图形界面制作完成后，要想让每一个组件都发挥自己的作用，就必须对所有的组件进行事件处理，才能实现软件与用户的交互。常用的事件有窗体事件、动作事件、焦点事件、鼠标事件和键盘事件。在 Swing 编程中，依然使用最早的 AWT 事件处理方式。下面先了解一下事件和监听器的概念。

10.8.1 事件和监听器

事件就是表示一个对象发生了状态变化。例如，一个按钮被单击，按钮的状态就发生了变化，此时就会产生一个事件。如果想处理此事件，就需要事件的监听者不断地监听事件的变化，并根据这些事件进行相应地处理。Swing 使用的是基于代理的事件模型。

基于代理（授权）事件模型，事件处理是一个事件源授权到一个或多个事件监听器，其基本原理是组件激发事件、事件监听器监听和处理事件，可以调用组件的 add<EventType>Listener 方法向组件注册监听器。将其加入到组件之后，如果组件激发了相应类型的事件，那么定义在监听器中的事件处理方法会被调用。

此模型主要由以三种对象为中心组成：

- 事件源：由它来激发产生事件，是产生或抛出事件的对象。
- 事件监听器：由它来处理事件，实现某个特定 EventListener 接口，此接口定义了一种或多种方法，事件源调用它们以响应该接口所处理的每一种特定事件类型。
- 事件：具体的事件类型，事件类型封装在以 java.util.EventObject 为根的类层次中。当事件发生时，事件记录发生的一切事件，并从事件源传播到监听器对象

如图 10-14 所示为事件的体系结构。

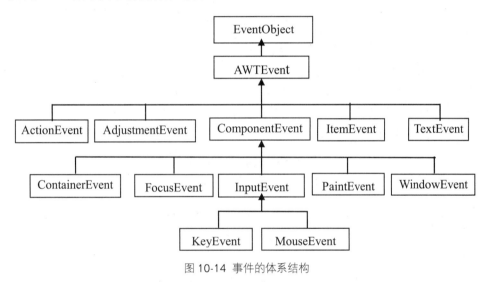

图 10-14 事件的体系结构

Java 事件的处理流程如图 10-15 所示。

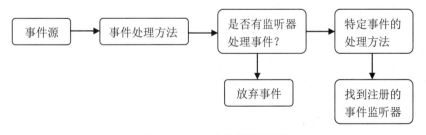

图 10-15 Java 事件的处理流程

Java 常用的事件类型及说明如表 10-9 所示。

表10-9 Java常用的事件类型及说明

事件类	说明	事件源
WindowEvent	当一个窗口激活、关闭、失效、恢复、最小化、打开或退出时会生成此事件	Windows
ActionEvent	通常在单击按钮、双击列表项或选中一个菜单项时生成此事件	Button、MenuItem、TextField、List
AdjustmentEvent	操纵滚动条时会生成此事件	Scrollbar
ComponentEvent	当一个组件移动、隐藏、调整大小或成为可见时会生成此事件	Component
ItemEvent	选中复选框或列表项时，或者当一个选择框或一个可选菜单的项被选择或取消时生成此事件	Checkbox、Choice、List、CheckboxMenuItem
FocusEvent	组件获得或失去焦点时会生成此事件	Component
KeyEvent	接收到键盘输入时会生成此事件	Component
MouseEvent	拖动、移动、单击、按下或释放鼠标，或者在鼠标进入或退出一个组件时生成此事件	Component
ContainerEvent	将组件添加至容器或从中删除时会生成此事件	Container
TextEvent	在文本区或文本域的文本改变时会生成此事件	TextField、TextArea

监听器通过实现 java.awt.event 包中定义的一个或多个接口来创建。在发生事件时，事件源将调用监听器定义的相应方法，有兴趣接收事件的任何监听器类都必须实现监听器接口。Java 的监听器接口如表 10-10 所示。

表10-10 Java的监听器接口

事件监听器	方法
ActionListener	actionPerformed
AdjustmentListener	adjustmentValueChanged
ComponentListener	componentResized、componentMoved、componentShown、componentHidden
ContainerListener	componentAdded、componentRemoved
FocusListener	focusLost、focusGained
ItemListener	itemStateChanged
KeyListener	keyPressed、keyReleased、keyTyped
MouseListener	mouseClicked、mouseEntered、mouseExited、mousePressed、mouseReleased
MouseMotionListener	mouseDragged、mouseMoved
TextListener	textChanged
WindowListener	windowActivated、windowDeactivated、windowClosed、windowClosing、windowIconified、windowDeiconified、windowOpened

10.8.2 窗体事件

WindowListener 是专门处理窗体的时间监听器接口,一个窗体的所有变化,如窗口的打开、关闭等都可以使用这个接口进行监听。WindowListener 接口定义的方法如表 10-11 所示。

表10-11 WindowListener接口定义的方法

序号	方　法	描　述
1	void windowOpened(WindowEvent e)	窗口打开时触发
2	void windowClosing(WindowEvent e)	当窗口正在关闭时触发
3	void windowClosed(WindowEvent e)	当窗口被关闭时触发
4	void windowIconified(WindowEvent e)	窗口最小化时触发
5	void windowDeiconified(WindowEvent e)	窗口从最小化恢复到正常状态时触发
6	void windowActivated(WindowEvent e)	窗口变为活动窗口时触发
7	void windowDeactivated(WindowEvent e)	将窗口变为不活动的窗口时触发

下面是实现 WindowListener 接口的示例。

```java
import java.awt.event.WindowEvent;
import java.awt.event.WindowListener;
public class MyWindowEventHandle implements WindowListener {
    @Override
    public void windowActivated(WindowEvent e) {
        System.out.println("windowActivated=== 窗口被选中！");
    }
    @Override
    public void windowClosed(WindowEvent e) {
        System.out.println("windowClosed=== 窗口被关闭！");
    }
    @Override
    public void windowClosing(WindowEvent e) {
        System.out.println("windowClosing=== 窗口关闭！");
    }
    @Override
    public void windowDeactivated(WindowEvent e) {
        System.out.println("indowDeactivated=== 取消窗口选中！");
    }
    @Override
    public void windowDeiconified(WindowEvent e) {
        System.out.println("windowDeiconified=== 窗口从最小化恢复！");
    }
    @Override
    public void windowIconified(WindowEvent e) {
        System.out.println("windowIconified=== 窗口最小化！");
    }
```

```
        @Override
        public void windowOpened(WindowEvent e) {
            System.out.println("windowOpened=== 窗口被打开！ ");
        }
    }
```

只有一个监听器是不够的，还需要在组件使用时注册监听，这样才可以处理动。直接使用窗体的 addWindowListener（监听对象）方法即可注册事件监听，如下面示例：

```
import java.awt.Color;
import javax.swing.JFrame;
public class MyWindowEventJFrameDemo {
    public static void main(String[] args) {
        JFrame f = new JFrame("WindowListener 示例 ");// 实例化窗体对象
        // 将此窗体加入到以下窗口事件监听器中，监听器就可以根据事件进行处理
        f.addWindowListener(new MyWindowEventHandle());
        f.setSize(300, 160);                    // 设置组件大小
        f.setBackground(Color.WHITE);           // 设置窗体的背景颜色
        f.setLocation(300, 200);                // 设置窗体的显示位置
        f.setVisible(true);                     // 让组件显示
    }
}
```

运行程序，会显示一个窗体，对窗体状态进行改变，则在后台会显示对应窗体操作的相类信息。一般在关闭监听 windowClosing 中编写 System.exit(1) 语句，这样关闭按钮就会真正起作用，可以让程序正常结束并退出。

```
Console  Problems  @ Javadoc  Decla
MyWindowEventJFrameDemo [Java Application] D:\Java
windowActivated===窗口被选中！
windowOpened===窗口被打开！
windowIconified===窗口最小化！
indowDeactivated===取消窗口选中！
windowDeiconified===窗口从最小化恢复！
windowActivated===窗口被选中！
windowClosing===窗口关闭！
indowDeactivated===取消窗口选中！
```

上面的示例在实现 WindowListener 接口时要实现接口的所有方法，但是这些方法在开发时不一定都要用到，那么就没有必要覆写那么多的方法，而是可以根据个人需要进行覆写。Java 在实现类和接口之间增加了一个过渡的抽象类，子类（适配器 Adapter 类）继承抽象类就可以根据自己的需要覆写方法，方便用户进行事件处理的实现。WindowListener 接口的适配器类是 WindowAdapter。

下面的示例是直接通过适配器类实现窗口的关闭。

```java
import java.awt.event.WindowAdapter;
import java.awt.event.WindowEvent;
public class MyWindowCloseDemo extends WindowAdapter {
    @Override
    public void windowClosing(WindowEvent e) {
            // 此类只覆写windowClosing()方法
            System.out.println("windowClosing=== 窗口关闭！");
            System.exit(1);// 系统退出
    }
}
```

在窗体的操作代码中直接使用上面的监听器类即可实现。代码如下：

```java
JFrame f = new JFrame("WindowClosing 示例 ");        // 实例化窗体对象
    // 直接使用WindowAdapter的子类完成监听处理
    f.addWindowListener(new MyWindowCloseDemo());
```

在程序开发中，往往采用匿名内部类来完成监听的操作，以减少监听器类的定义。

下面的示例演示了匿名内部类的使用方法，建议读者掌握和使用这种方法。

```java
import java.awt.Color;
import java.awt.event.WindowAdapter;
import java.awt.event.WindowEvent;
import javax.swing.JFrame;
public class TestWindowClose {
    public static void main(String[] args) {
            JFrame f = new JFrame("WindowClosing 示例 ");// 实例化窗体对象
            // 直接使用WindowAdapter的子类完成监听处理
            f.addWindowListener(new WindowAdapter(){
                // 覆写窗口关闭的方法
                @Override
                public void windowClosing(WindowEvent e) {
                    System.exit(1);
                }
            });
            f.setSize(300, 160);                    // 设置组件大小
            f.setBackground(Color.WHITE);           // 设置窗体的背景颜色
            f.setLocation(300, 200);                // 设置窗体的显示位置
            f.setVisible(true);                     // 让组件显示
    }
}
```

Java常用的适配器类如表10-12所示。

表10-12 Java常用的适配器类

适配器类	事件监听器接口
ComponentAdapter	ComponentListener
ContainerAdapter	ContainerListener
FocusAdapter	FocusListener
KeyAdapter	KeyListener
MouseAdapter	MouseListener
MouseMotionAdapter	MouseMotionListener
WindowAdapter	WindowListener

10.8.3 动作事件及监听处理

Swing 动作事件由 ActionEvent 类捕获，常用的是当单击按钮后将发出动作事件，可以通过实现 ActionLinstener 接口处理相应的动作事件。

ActionLinstener 接口只有一个抽象方法，当动作发生后被触发。具体定义如下：

```java
public interface ActionListener extends EventListener {
    public void actionPerformed(ActionEvent e);
}
```

ActionEvent 类中有两个常用的方法：

- public Object getSource()：用来获得触发此次事件的组件对象。
- public String getActionCommand()：用来获得与当前动作相关的命令字符串。

下面的示例演示了使用以上监听接口监听按钮的单击事件。

```java
import java.awt.BorderLayout;
import java.awt.event.ActionEvent;
import java.awt.event.ActionListener;
import javax.swing.JButton;
import javax.swing.JFrame;
import javax.swing.JLabel;
public class ActionEventDemo{
    private JLabel lb;              // 声明一个标签对象，用于显示提示信息
    private JButton b;              // 声明一个按钮对象
    ActionEventDemo(){
        JFrame f = new JFrame("演示");
        lb = new JLabel("欢迎登录！");
        lb.setHorizontalAlignment(JLabel.CENTER);
        b = new JButton("登录");
        b.addActionListener(new ActionListener(){
            public void actionPerformed(ActionEvent e) {
```

```
                    JButton button = (JButton)e.getSource();
                    String buttonName = e.getActionCommand();
                    if (buttonName.equals("登录")) {
                        lb.setText("您已经成功登录！");
                        button.setText("退出");
                    } else {
                        lb.setText("您已经安全退出！");
                        button.setText("登录");
                    }
                }
            });
            f.add(lb);
            f.add(b,BorderLayout.SOUTH);
            f.setBounds(100, 100, 230, 120);
            f.setLocation(100, 80);
            f.setVisible(true);
            f.setDefaultCloseOperation(JFrame.EXIT_ON_CLOSE);
        }
        public static void main(String[] args) {
            new ActionEventDemo();
        }
    }
```

程序运行结果如图 10-16 所示。

初次运行时的效果

单击"登录"按钮后

单击"退出"按钮后

图 10-16 程序运行结果

10.8.4 键盘事件及监听处理

键盘事件由 KeyEvent 类捕获，常用的是当向文本框输入内容时将触发键盘事件，可以通过 KeyListener 接口处理相应的键盘事件。有三个抽象方法，具体定义如下：

```
public interface KeyListener extends EventListener {
    // 输入某个键时调用
public void keyTyped(KeyEvent e);
// 键盘按下时调用
    public void keyPressed(KeyEvent e);
// 键盘松开时调用
```

```
    public void keyReleased(KeyEvent e);
}
```

如果要取得键盘输入的内容,就可以通过 KeyEvent 类捕获。KeyEvent 类的常用方法如表 10-13 所示。

表10-13 KeyEvent类的常用方法

序号	方法	描述
1	public char getKeyChar()	返回输入的字符,只针对keyTyped有意义
2	public void setKeyChar(char keyChar)	返回输入字符的键码
3	public static String getKeyText(int keyCode)	返回此键的信息,如"F3""A"等

下面的示例演示了使用 KeyAdapter 适配器完成键盘事件的监听。

```java
import java.awt.event.KeyAdapter;
import java.awt.event.KeyEvent;
import java.awt.event.WindowAdapter;
import java.awt.event.WindowEvent;
import javax.swing.JFrame;
import javax.swing.JScrollPane;
import javax.swing.JTextArea;
public class KeyEventDemo extends JFrame{
    private JTextArea text = new JTextArea();
    public KeyEventDemo(){
        super.setTitle("键盘事件");
        JScrollPane scroll = new JScrollPane(text);// 加入滚动条
        scroll.setBounds(5, 5, 300, 200);
        super.add(scroll);                          // 向窗体加入组件
        text.addKeyListener(new KeyAdapter(){
            public void keyTyped(KeyEvent e) {      // 输入内容
                text.append("输入的内容是: "+e.getKeyChar()+"\n");
            }
        });                                         // 加入key监听
        super.setSize(300, 200);
        super.setVisible(true);
        super.addWindowListener(new WindowAdapter(){
            public void windowClosing(WindowEvent e) {
                System.exit(1);
            }
        });
    }
    public static void main(String[] args) {
        new KeyEventDemo();
    }
}
```

}

程序运行结果如图 10-17 所示。

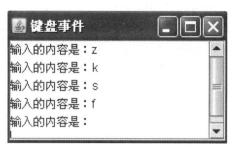

图 10-17　程序运行结果

10.8.5　鼠标事件及监听处理

鼠标事件是由 MouseEvent 类捕获，所有的组件都能产生鼠标事件，可以通过实现 MouseListener 接口处理相应的鼠标事件。

MouseListener 接口有 5 个抽象方法，具体定义如下：

```
public interface MouseListener extends EventListener {
    // 鼠标单击时调用（按下并释放）
public void mouseClicked(MouseEvent e);
// 鼠标按下时调用
    public void mousePressed(MouseEvent e);
// 鼠标释放时调用
    public void mouseReleased(MouseEvent e);
// 鼠标进入组件时调用
    public void mouseEntered(MouseEvent e);
// 鼠标离开组件时调用
    public void mouseExited(MouseEvent e);
}
```

每个事件触发后都会产生 MouseEvent 事件，此事件可以得到鼠标的相关操作。MouseEvent 类的常用方法及常量如表 10-14 所示。

表10-14　MouseEvent类的常用方法及常量

序号	方法及常量	类型	描述
1	public static final int BUTTON1	常量	表示鼠标左键的常量
2	public static final int BUTTON2	常量	表示鼠标滚轴的常量
3	public static final int BUTTON3	常量	表示鼠标右键的常量
4	public int getButton()	普通	以数字形式返回按下的鼠标键
5	public int getClickCount()	普通	返回鼠标的单击次数

下面的示例演示了使用 MouseAdapter 适配器完成鼠标事件的监听。

```java
import java.awt.event.MouseAdapter;
import java.awt.event.MouseEvent;
import java.awt.event.WindowAdapter;
import java.awt.event.WindowEvent;
import javax.swing.JFrame;
import javax.swing.JScrollPane;
import javax.swing.JTextArea;
public class MouseEventDemo extends JFrame{
    private JTextArea text = new JTextArea();
    public MouseEventDemo(){
        super.setTitle("鼠标事件");
        JScrollPane scroll = new JScrollPane(text);
        scroll.setBounds(5, 5, 300, 200);
        super.add(scroll);
        text.addMouseListener(new MouseAdapter(){
            public void mouseClicked(MouseEvent e) {
                int c = e.getButton();
                String mouseInfo = null;
                if (c == MouseEvent.BUTTON1) {
                    mouseInfo = "左键";
                } else if(c == MouseEvent.BUTTON2){
                    mouseInfo = "滚轴";
                } else {
                    mouseInfo = "右键";
                }
                text.append("鼠标单击: " + mouseInfo +"。\n");
            }
        });
        super.setSize(300, 200);
        super.setVisible(true);
        super.addWindowListener(new WindowAdapter(){
            public void windowClosing(WindowEvent e) {
                System.exit(1);
            }
        });
    }
    public static void main(String[] args) {
        new MouseEventDemo();
    }
}
```

程序运行结果如图 10-18 所示。

第 10 章　Java 图形用户界面

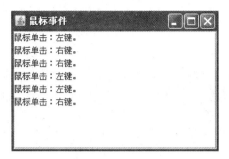

图 10-18　程序运行结果

10.8.6 焦点事件及监听处理

焦点事件是由 FocusEvent 类捕获，所有的组件都能产生焦点事件，可以通过实现 FocusListener 接口处理相应的焦点事件。FocusListener 接口有两个抽象方法，分别在组件获得或失去焦点时被触发，具体定义如下：

```
public interface FocusListener extends EventListener {
    // 组件获得焦点时触发该方法
    public void focusGained(FocusEvent e);
    // 组件失去焦点时触发该方法
    public void focusLost(FocusEvent e);
}
```

FocusEvent 类中常用的方法是 getSource()，用于获得触发此次事件的组件对象，返回值为 Object。

下面的示例演示了文本框获得焦点和失去焦点时的事件处理方法。

```
import java.awt.event.*;
import javax.swing.*;
public class FocusEventDemo {
    private JFrame  f = new JFrame(" 文本框的焦点事件 ");
    private JLabel lab = new JLabel("QQ 号码 ");
    private JTextField text = new JTextField(" 请输入 QQ 号码 ");
    private JLabel lab1 = new JLabel();
    public FocusEventDemo(){
        f.setLayout(null);
        lab.setBounds(30, 30, 60, 30);
        f.add(lab);
        text.setBounds(100, 30, 100, 30);
        text.addFocusListener(new FocusAdapter() {
            public void focusGained(FocusEvent e) {
                // 文本框获得焦点时清空文本框内容
                lab.setText("");
```

223

```
              }
              public void focusLost(FocusEvent e) {
                  // 文本框失去焦点时，在标签中显示文本框内容
                  lab1.setText(lab.getText());
              }
          });
          f.add(text);
          lab1.setBounds(60, 80, 100, 30);
          f.add(lab1);
          f.setSize(300, 200);
          f.setLocation(300, 200);
          f.setVisible(true);
          f.setDefaultCloseOperation(JFrame.EXIT_ON_CLOSE);
    }
    public static void main(String[] args) {
        new FocusEventDemo();
    }
}
```

10.9 单选按钮组件：JRadioButton

JRadioButton 组件是在给出的多个信息中指定一个。在 Swing 中可以使用 JRadioButton 组件完成一组单选按钮的操作。JRadioButton 类单独使用时，该单选按钮可以选中和取消选中；与 ButtonGroup 类联合使用时，则组成单选按钮组，用户只能选中按钮组中的一个单选按钮。取消选中的操作由 ButtonGroup 类自动完成。

ButtonGroup 类用于创建一个按钮组。按钮组的作用是负责维护该按钮组的"开启"状态，在按钮组中只能有一个按钮处于"开启"状态。按钮组经常用来维护由 JRadioButton、JRadioButtonMenuItem 或 JToggleButton 类型的按钮组成的按钮组。ButtonGroup 类的常用方法如表 10-15 所示。

表10-15 ButtonGroup类的常用方法

序号	方法	描述
1	public void add(AbstractButton b)	将按钮添加到按钮组中
2	public void remove(AbstractButton b)	从按钮组中移除按钮
3	public int getButtonCount()	取得按钮组中按钮的个数
4	public Enumeration<AbstractButton> getElements()	取得按钮组中所有按钮

JRadioButton 类的常用方法如表 10-16 所示。

第 10 章　Java 图形用户界面

表10-16　JRadioButton类的常用方法

序号	方法	描　述
1	public JRadioButton(Icon icon)	将按钮添加到按钮组中
2	public JRadioButton(Icon icon, boolean selected)	从按钮组中移除按钮
3	public JRadioButton(String text)	取得按钮组中按钮的个数
4	public JRadioButton(String text, boolean selected)	取得按钮组中所有按钮
5	public void setIcon(Icon defaultIcon)	设置图片
6	public void setText(String text)	设置显示文本
7	public void setSelected(boolean b)	设置是否选中
8	public boolean isSelected()	返回是否被选中

下面的示例演示了 JRadioButton 类和 ButtonGroup 类的使用方法。

```
import java.awt.*;
import javax.swing.*;
public class JradioButtonDemo01 extends JFrame{
    private JLabel l = new JLabel("请选择你的职业：");
    private JRadioButton rb1 = new JRadioButton("公务员");
    private JRadioButton rb2 = new JRadioButton("教师");
    private JRadioButton rb3 = new JRadioButton("工人");
    private ButtonGroup bg = new ButtonGroup();
    private JPanel p = new JPanel();
    public JRadioButtonDemo(){
            setTitle("JRadioButton演示");
            setLayout(new GridLayout(1,4));
            getContentPane().add(l);
            bg.add(rb1);
            getContentPane().add(rb1);
            bg.add(rb2);
            getContentPane().add(rb2);
            rb2.setSelected(true);
            bg.add(rb3);
            getContentPane().add(rb3);
            setBounds(100, 100, 180, 90);
            setLocation(300, 80);
            setVisible(true);
            setDefaultCloseOperation(JFrame.EXIT_ON_CLOSE);
    }
    public static void main(String[] args) {
            new JRadioButtonDemo();
    }
}
```

程序运行结果如图 10-19 所示。

图 10-19 程序运行结果

JRadioButton 的事件处理使用 ItemListener 接口进行事件监听。

```
import java.awt.*;
import java.awt.event.*;
import javax.swing.*;
public class JradioButtonDemo02 extends JFrame{
    private JLabel l = new JLabel("请选择你的职业：");
    private JRadioButton rb1 = new JRadioButton(" 公务员 ");
    private JRadioButton rb2 = new JRadioButton(" 教师 ");
    private JRadioButton rb3 = new JRadioButton(" 工人 ");
    private ButtonGroup bg = new ButtonGroup();
    private JPanel p = new JPanel();
    private JLabel l2 = new JLabel();
    public JRadioButtonDemo(){
        setTitle("JRadioButton 演示 ");
        setLayout(new GridLayout(2,3));
        getContentPane().add(l);
        bg.add(rb1);
        rb1.addItemListener(new ItemListener() {
            public void itemStateChanged(ItemEvent e) {
                changed(e);
            }
        });
        getContentPane().add(rb1);
        bg.add(rb2);
        rb2.setSelected(true);
        rb2.addItemListener(new ItemListener() {
            public void itemStateChanged(ItemEvent e) {
                changed(e);
            }
        });
        getContentPane().add(rb2);
        bg.add(rb3);
        rb3.addItemListener(new ItemListener() {
            public void itemStateChanged(ItemEvent e) {
                changed(e);
            }
        });
```

```
            getContentPane().add(rb3);
            getContentPane().add(l2);
            setBounds(100, 100, 430, 90);
            setLocation(300, 80);
            setVisible(true);
            setDefaultCloseOperation(JFrame.EXIT_ON_CLOSE);
        }
        public void changed(ItemEvent e) {
            if (e.getSource() == rb1) {
                l2.setText("你的职业是公务员！");
            } else if(e.getSource() == rb2){
                l2.setText("你的职业是教师！");
            } else {
                l2.setText("你的职业是工人！");
            }
        }
        public static void main(String[] args) {
            new JRadioButtonDemo();
        }
}
```

程序运行结果如图 10-20 所示。

图 10-20 程序运行结果

10.10 复选框组件：JCheckBox

JCheckBox 组件实现一个复选框，该复选框可以被选中和取消选中，并且可以同时选中多个。常用的方法是 JCheckBox 类的构造方法。JCheckBox 和 JRadioButton 的事件处理监听接口是一样的，即都有 ItemListener 接口。下面的示例演示了复选框的使用及事件处理。

```
import java.awt.*;
import java.awt.event.*;
import javax.swing.*;
public class JCheckBoxDemo {
    private JFrame f = new JFrame("JCheckBox演示");
    private Container c = f.getContentPane();
    private JCheckBox cb1 = new JCheckBox("篮球");
    private JCheckBox cb2 = new JCheckBox("游泳");
```

```java
    private JCheckBox cb3 = new JCheckBox(" 跑步 ");
    JLabel l = new JLabel(" 你喜欢的运动："); 
    JPanel p = new JPanel();
    public JCheckBoxDemo02(){
        // 定义一个边框显示条
        p.setBorder(BorderFactory.createTitledBorder(" 你喜欢的运动 "));
        p.setLayout(new GridLayout(2,3));
        cb1.addItemListener(new ItemListener() {
            public void itemStateChanged(ItemEvent e) {
                selected1(e);
            }
        });
        p.add(cb1);
        cb2.addItemListener(new ItemListener() {
            public void itemStateChanged(ItemEvent e) {
                selected2(e);
            }
        });
        p.add(cb2);
        cb3.addItemListener(new ItemListener() {
            public void itemStateChanged(ItemEvent e) {
                selected3(e);
            }
        });
        p.add(cb3);
        p.add(l);
        c.add(p);
        f.setSize(360, 100);
        f.setVisible(true);
        f.setDefaultCloseOperation(JFrame.EXIT_ON_CLOSE);
    }
    public void selected1(ItemEvent e){
        String sb = l.getText();
        if (cb1.isSelected()) {
            l.setText(l.getText()+cb1.getText());
        } else{
            l.setText(sb.replaceAll(" 篮球 ", ""));
        }
    }
    public void selected2(ItemEvent e){
        String sb = l.getText();
        if (cb2.isSelected()) {
            l.setText(l.getText()+cb2.getText());
        } else{
```

```
                l.setText(sb.replaceAll("游泳", ""));
            }
    }
    public void selected3(ItemEvent e){
            String sb = l.getText();
            if (cb3.isSelected()) {
                    l.setText(l.getText()+cb3.getText());
            } else{
                    l.setText(sb.replaceAll("跑步", ""));
            }
    }
    public static void main(String[] args) {
            new JCheckBoxDemo02();
    }
}
```

程序运行结果如图 10-21 所示。

图 10-21 程序运行结果

10.11 列表框组件：JList

列表框可以同时将多个选项信息以列表的方式展现给用户，使用 JList 可以构建一个列表框。在 JList 的构造方法中经常使用 ListModel 构造 JList 对象。ListModel 是一个专门用于创建 JList 列表内容的接口。

JList 使用 ListSelectionListener 的监听接口实现对 JList 中的选项进行监听，此接口的定义如下：

```
public interface ListSelectionListener extends EventListener{
 // 当值发生改变触发
 void valueChanged(ListSelectionEvent e);
}
```

下面的示例演示了列表框组件的创建及事件处理。

```
import java.awt.*;
import javax.swing.*;
import javax.swing.event.*;
```

```java
class MyListModel extends AbstractListModel{
    private String[] inst = {"篮球","排球","足球","乒乓球","网球"};
    public Object getElementAt(int index) {
        if (index < this.inst.length) {
            return inst[index];
        } else {
            return null;
        }
    }
    public int getSize() {
        return this.inst.length;
    }
}
class MyList implements ListSelectionListener{
    private JFrame f = new JFrame("JList 演示");
    private Container c = f.getContentPane();
    private JList list = null;
    public MyList(){
        list = new JList(new MyListModel());
        list.setBorder(BorderFactory.createTitledBorder(" 你喜欢的球类运动 "));
        list.addListSelectionListener(this);
        c.add(new JScrollPane(this.list));
        f.setSize(300, 180);
        f.setVisible(true);
        f.setDefaultCloseOperation(JFrame.EXIT_ON_CLOSE);
    }
    public void valueChanged(ListSelectionEvent e) {
        int[] temp = list.getSelectedIndices();
        System.out.print("选定的内容：");
        for (int i = 0; i < temp.length; i++) {
            System.out.print(list.getModel().getElementAt(i)+"、");
        }
        System.out.println();
    }
}
public class JListDemo {
    public static void main(String[] args) {
        new MyList();
    }}
```

10.12 下拉列表框：JComboBox

JComboBox 下拉列表框既有列表，又可以自己输入数据，它的事件监听接口是 ItemListener。

下面的示例演示了 JComboBox 下拉列表框的使用及事件处理。

```java
import java.awt.*;
import java.awt.event.*;
import javax.swing.*;
public class JComboBoxDemo {
    private JLabel label,label2;
    public JComboBoxDemo(){
        JFrame f = new JFrame();
        Container c = f.getContentPane();
        f.setLayout(null);
        f.setSize(300, 200);
        f.setVisible(true);
        f.setTitle("选择框测试 ");
        f.setDefaultCloseOperation(f.EXIT_ON_CLOSE);
        label = new JLabel(" 学历:");
        label.setBounds(20, 35, 60, 20);
        f.add(label);
        String[] schooRecord = { "本科", "硕士", "博士" };
        JComboBox comboBox = new JComboBox(schooRecord);
        comboBox.setBounds(85, 35, 100, 20);
        comboBox.setEditable(true);
        comboBox.setMaximumRowCount(4);
        f.add(comboBox);
        comboBox.insertItemAt("大专", 0);
        comboBox.setSelectedIndex(3);
        comboBox.addItem("专升本");
        label2 = new JLabel();
        label2.setBounds(200, 35, 100, 20);
        f.add(label2);
        comboBox.addItemListener(new ItemListener() {
            public void itemStateChanged(ItemEvent e) {
                if (e.getStateChange() == ItemEvent.SELECTED) {
                    String itemSize = (String)e.getItem();
                    label2.setText(itemSize);
                }
            }
        });
    }
    public static void main(String[] args) {
        new JComboBoxDemo();
    }
}
```

程序运行结果如图 10-22 所示。

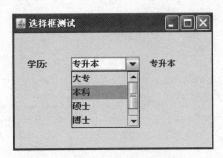

图 10-22 程序运行结果

10.13 菜单组件：JMenu 与 JMenuBar

JMenuBar 组件的功能是用来摆放 JMenu 组件。当建立完许多的 JMenu 组件后，需要通过 JMenuBar 组件将 JMenu 组件加入到窗口中。

下面通过 JMenu 与 JMenuBar 组件构建一个简单的菜单。

```java
import javax.swing.*;
public class JMenuDemo {
    public static void main(String[] args) {
        JFrame f = new JFrame("菜单演示");// 定义窗体
        JTextArea ta = new JTextArea();   // 定义文本框
        ta.setEditable(true);             // 定义文本框可以编辑
        // 在面板中加入文本框及滚动条
        f.getContentPane().add(new JScrollPane(ta));
        JMenu menuFile = new JMenu("文件"); // 定义 JMenu 组件—创建菜单
        JMenuBar menuBar = new JMenuBar();  // 定义 JMenuBar 组件
        JMenuItem newItem = new JMenuItem("新建");   // 创建菜单项
        newItem.setMnemonic('N');          // 设置快捷键
        JMenuItem openItem = new JMenuItem("打开");  // 创建菜单项
        openItem.setMnemonic('O');         // 设置快捷键
        JMenuItem saveItem = new JMenuItem("保存");  // 创建菜单项
        saveItem.setMnemonic('S');         // 设置快捷键
        JMenuItem exitItem = new JMenuItem("退出");  // 创建菜单项
        exitItem.setMnemonic('E');         // 设置快捷键
        menuFile.add(newItem);             // 加入菜单项
        menuFile.add(openItem);            // 加入菜单项
        menuFile.add(saveItem);            // 加入菜单项
        menuFile.addSeparator();           // 加入分割线
        menuFile.add(exitItem);            // 加入菜单项
        menuBar.add(menuFile);             // 加入 JMenu
```

```
        f.add(menuBar);                      // 窗体中加入 JMenuBar 组件
        f.setSize(300, 180);
        f.setLocation(300, 200);
        f.setVisible(true);
    }
}
```

JMenuItem 与 JButton 的事件处理机制是完全一样的，即选择一个菜单项与单击一个按钮的效果是完全相同的，这里不再赘述。

10.14 文件选择框组件：JFileChooser

在处理窗口上的一些操作时，如希望将一个文本编辑器上的文字保存起来，供以后方便使用，系统应当提供一个存储文件的对话框，将文字保存到一个自定义或内定的文件名中，或者选择想要打开的文件，在 java 中，这些操作都可以由 JFileChooser 组件来完成。JFileChooser 组件不仅提供了打开文件存盘的窗口功能，还提供了显示特定类型文件图标的功能，亦能针对某些文件类型做过滤操作。

下面的示例演示了在打开文件、编辑文件和保存文件的过程中 FileChooser 组件的使用方法。

```java
import javax.swing.*;
import java.awt.*;
import java.awt.event.*;
import java.io.*;
class FileChooserDemo1 implements ActionListener {
    JFrame f = null;
    JLabel label = null;
    JTextArea textarea = null;
    JFileChooser fileChooser = null;
    public FileChooserDemo1() {
        f = new JFrame("文件选择框演示");
        Container contentPane = f.getContentPane();
        textarea = new JTextArea();
        JScrollPane scrollPane = new JScrollPane(textarea);
        scrollPane.setPreferredSize(new Dimension(350, 300));
        JPanel panel = new JPanel();
        JButton b1 = new JButton("打开文件");
        b1.addActionListener(this);
        JButton b2 = new JButton("保存文件");
        b2.addActionListener(this);
        panel.add(b1);
```

```java
            panel.add(b2);
            label = new JLabel(" ", JLabel.CENTER);
            // 建立一个FileChooser对象，并指定D:的目录为默认文件对话框路径.
            fileChooser = new JFileChooser("D:\\");
            contentPane.add(label, BorderLayout.NORTH);
            contentPane.add(scrollPane, BorderLayout.CENTER);
            contentPane.add(panel, BorderLayout.SOUTH);
            f.pack();
            f.setVisible(true);
            f.setDefaultCloseOperation(f.EXIT_ON_CLOSE);
    }
    public static void main(String[] args) {
            new FileChooserDemo1();
    }
    public void actionPerformed(ActionEvent e) {
            File file = null;
            int result;
            if (e.getActionCommand().equals("打开文件")) {
                    fileChooser.setApproveButtonText("确定");
                    fileChooser.setDialogTitle("打开文件");
                    result = fileChooser.showOpenDialog(f);
                    textarea.setText("");
                    if (result == JFileChooser.APPROVE_OPTION) {
                            file = fileChooser.getSelectedFile();
                            label.setText("您选择打开的文件名称为: " +
file.getName());
                    } else if (result == JFileChooser.CANCEL_OPTION) {
                            label.setText("您没有选择任何文件");
                    }
                    FileInputStream fileInStream = null;
                    if (file != null) {
                            try {
                                    fileInStream = new FileInputStream(file);
                            } catch (FileNotFoundException fe) {
                                    label.setText("File Not Found");
                                    return;
                            }
                            int readbyte;
                            try {
                                    while ((readbyte = fileInStream.read()) != -1) {
                                    textarea.append(String.valueOf((char) readbyte));
                                    }
                            } catch (IOException ioe) {
                                    label.setText("读取文件错误");
```

```java
            } finally {// 回收 FileInputStream 对象，避免资源浪费
                try {
                    if (fileInStream != null)
                        fileInStream.close();
                } catch (IOException ioe2) {}
            }
        }
    }
    // 实作写入文件的功能
    if (e.getActionCommand().equals("存储文件")) {
        result = fileChooser.showSaveDialog(f);
        file = null;
        String fileName;
        if (result == JFileChooser.APPROVE_OPTION) {
            file = fileChooser.getSelectedFile();
            label.setText("您选择存储的文件名称为： " +
file.getName());
        } else if (result == JFileChooser.CANCEL_OPTION) {
            label.setText("您没有选择任何文件");
        }
        FileOutputStream fileOutStream = null;
        if (file != null) {
            try {
                fileOutStream = new FileOutputStream(file);
            } catch (FileNotFoundException fe) {
                label.setText("File Not Found");
                return;
            }
            String content = textarea.getText();
            try {
                fileOutStream.write(content.getBytes());
            } catch (IOException ioe) {
                label.setText("写入文件错误");
            } finally {
                try {
                    if (fileOutStream != null)
                        fileOutStream.close();
                } catch (IOException ioe2) {}
            }
        }
    }
}
}
```

程序运行结果如图 10-23 所示。

图 10-23 程序运行结果

程序执行时，通过单击"打开文件"按钮触发一个事件，打开一个文件选择框；通过单击"存储文件"按钮触发一个事件，打开一个保存文件的选择框。

10.15 要点总结

本章简单介绍了 AWT 和 Swing 的概念，了解了组件、容器和布局管理器。重点介绍了常用的组件、容器和事件处理机制，通过捕获组件的各种事件，即可完成相应的业务逻辑。

10.16 练习题

1. 填空题

（1）Java 图形用户界面的基本组成部分是组件。组件是以 _____ 显示在屏幕上能进行 _____ 的对象。

（2）容器是一种特殊的 _____，其主要功能是容纳其他组件和容器。

（3）通过布局管理器可以使我们生成的图形用户界面具有良好的 _____。

（4）Component 类直接继承自 _____ 类，是一个抽象类。

（5）Frame 类用于建立标准的窗口，继承自 _____ 类。

（6）AWT 基于本地对等组件的 _____ 体系结构，而 Swing 是由纯 Java 实现的，其组件不依赖于 _____。

（7）Swing 的组件几乎都是 _____，而 AWT 的组件被称作 _____。

（8）Swing 不但在 _____ 上表现一致，而且还提供了 _____ 不支持的特性。

（9）Swing 采用 MVC（Model-View-Controller）设计模式，即 _____。

（10）Java AWT 事件处理模型将 _____ 和 _____ 完全分开。

（11）Java 的事件处理采用 _____ 模式，即事件源可以把在其自身所有可能发生的事件分别授权给 _____。

（12）事件处理者一直监听 _____ 上发生的事件，一旦发现是其 _____ 就马上进行处理。这也是事件处理者被称作监听器的原因。

（13）除了实现事件监听器接口外，还可以通过 _____ 构造监听器。Java 语言为一些监听器接口提供了 _____。

（14）AWT 事件可以分为两大类，即 _____ 和 _____。

2. 选择题

（1）下面没有被 Component 类实现的接口是 _____。

  A. ImageObserver B. MenuContainer C. Serializable D. Clone

（2）Frame 类默认的布局管理器是 _____。

  A. BorderLayout B. FlowLayout C. CardLayout D.GridLayout

（3）Frame 类直接继承自下面哪个类 _____。

  A. Container B. Window C. Component D. Object

（4）下列哪个布局管理器会把加入的组件像卡片一样重叠放置，使用者第一次只能看到最上面的卡片 _____。

  A. BorderLayout B. FlowLayout C. CardLayout D.GridLayout

（5）GridBagLayout 布局管理器不限定加入组件的大小都相同，通过下面哪个类设置每个组件的大小 _____。

  A. GridBagConstraints B. GridLayout C. Frame D. Window

（6）下面 Swing 类中，没有继承 JComponent 类的是 _____。

  A. JFrame B. JComboBox C. JTable D. JTree

（7）下面 Swing 类中，没有实现 Accessible 接口的是 _____。

  A. JTree B. JTabbedPane C. JOptionPane D. JComponent

（8）在下列 Swing 组件中，哪个组件可以用于构建对话框 _____。

  A. JInternalFrame B. JOptionPane C. JFrame D. JTabbedPane

（9）在下列 Swing 组件中，哪个组件可以添加标签页 _____。

A. JInternalFrame   B. JOptionPane   C. JFrame   D. JTabbedPane

（10）在下列 Swing 组件中，哪个组件可以用来分隔窗体_____。

A. JComponent   B. JSplitPane   C. JFrame   D. JTabbedPane

（11）与 AWT 有关的所有事件类都继承自 AWTEvent 类。下列哪个类不是继承自 AWTEvent 类_____。

A. EventObject   B. KeyEvent   C. ComponentEvent   D. MouseEvent

（12）下列事件中不属于低级事件的是_____。

A. ContainerEvent   B. ActionEvent   C. MouseEvent   D. FocusEvent

（13）下列事件中不属于高级事件的是_____。

A. AdjustmentEvent   B. ItemEvent   C. ComponentEvent   D. TextEvent

（14）窗口被关闭触发的事件被封装在下列哪个类中_____。

A. WindowEvent   B. AdjustmentEvent   C. ItemEvent   D. TextEvent

（15）滑动滚动条触发的事件被封装在下列哪个类中_____。

A. WindowEvent   B. AdjustmentEvent   C. ItemEvent   D. TextEvent

第 11 章
Java 常用类库

　　Java 的应用程序接口（API）以包的形式进行组织，每个包提供了大量的相关类、接口和异常处理类，这些包的集合就称为 Java 类库。本章介绍 Java 常用类，包括 StringBuffer 类、Runtime 类、System 类、Math 类及 Random 类。

11.1 StringBuffer 类

　　StringBuffer 类支持的方法大部分与 String 类似，因为 StringBuffer 类在开发中可以提升代码的性能，所以使用较多。Java 为了保证用户操作的适应性，在 StringBuffer 类中定义的方法名称大部分都与 String 一样，读者可自行查询 JDK 文档。下面介绍一些常用的方法。

1. 字符串连接操作

　　在程序中使用 append() 方法可以进行字符串的连接，而且此方法返回了一个 StringBuffer 类的实例。这样可以使用代码链的形式一直调用 append() 方法，代码如下：

```
public class StringBufferDemo01 {
    public static void main(String[] args) {
        StringBuffer sb = new StringBuffer();              // 声明对象
        sb.append("zknu.");                                // 向 StringBuffer 中添加内容
        sb.append("edu.").append("cn");                    // 连续调用 append 方法添加内容
        sb.append("\n");                                   // 添加一个转义符表示换行
        sb.append(" 数字 = ").append(3).append("\n");     // 添加数字
```

```
            sb.append(" 字符 = ").append('c').append("\n");// 添加字符
            sb.append(" 布尔 = ").append(false);          // 添加布尔类型
            System.out.println(sb);                       // 内容输出
    }
}
```

程序运行结果如图 11-1 所示。

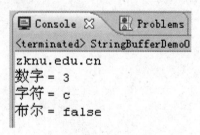

图 11-1 程序运行结果

2. 在指定位置添加内容

可以直接使用 insert() 方法在指定位置为 StringBuffer 添加内容，代码如下：

```
public class StringBufferDemo02 {
    public static void main(String[] args) {
        StringBuffer sb = new StringBuffer();       // 声明对象
        sb.append(" 计算机科学系 ");                  // 向 StringBuffer 中添加内容
        sb.insert(0, " 周口师范学院 ");               // 在所有内容之前添加
        System.out.println(sb);                     // 内容输出
        sb.insert(sb.length(), "- 陈占伟 ");         // 在最后添加
        System.out.println(sb);                     // 内容输出
    }
}
```

程序运行结果如图 11-2 所示。

图 11-2 程序运行结果

3. 字符串反转操作

字符串反转是一种常见的操作，较早的字符串反转是由栈来完成的。在 StringBuffer 类中专门提供了字符串反转的操作方法，代码如下：

```
public class StringBufferDemo03 {
```

```java
    public static void main(String[] args) {
        StringBuffer sb = new StringBuffer();   // 声明对象
        sb.append("计算机科学系");               // 向 StringBuffer 中添加内容
        sb.insert(0, "周口师范学院");            // 在所有内容之前添加
        // 将内容反转后变为 String 类型
        String s = sb.reverse().toString();
        System.out.println(s);                  // 内容输出
    }
}
```

程序运行结果如图 11-3 所示。

图 11-3 程序运行结果

4. 替换指定范围的内容

在 StringBuffer 类中也存在 replace() 方法，使用此方法可以对指定范围的内容进行替换。代码如下：

```java
public class StringBufferDemo04 {
    public static void main(String[] args) {
        StringBuffer sb = new StringBuffer();       // 声明对象
        sb.append("JKX->rjxy").append("SAM");       // 添加内容
        sb.replace(9, 12, "->czw");                 // 将 SAM 替换为 ->czw
        System.out.println(sb);                     // 内容输出
    }
}
```

程序运行结果如图 11-4 所示。

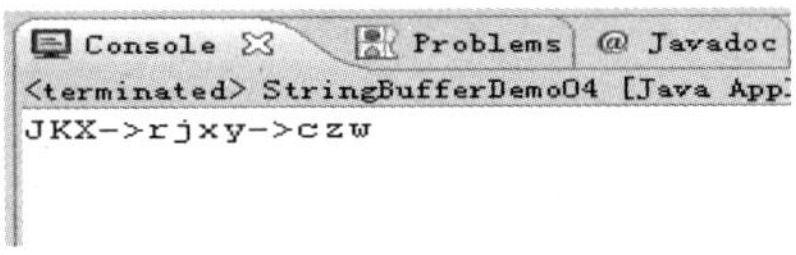

图 11-4 程序运行结果

在 String 中进行替换使用的是 replaceAll() 方法，读者需注意。

5. 删除指定范围的字符串

因为 StringBuffer 本身的内容是可以更改的，所以可以通过 delete() 方法删除指定范围的内容。代码如下：

```java
public class StringBufferDemo05 {
```

```java
    public static void main(String[] args) {
        StringBuffer sb = new StringBuffer();            // 声明对象
        sb.append("JKX->rjxy").append("SAM");            // 添加内容
        sb.replace(9, 12, "->czw");                      // 将 SAM 替换为 ->czw
        sb.delete(3, 9);                                 // 删除指定范围的字符串
        System.out.println(" 删除之后的内容: "+ sb);      // 内容输出
    }
}
```

程序运行结果如图 11-5 所示。

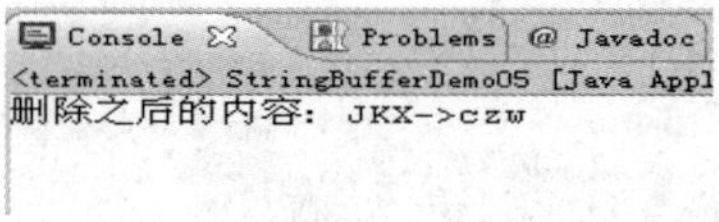

图 11-5 程序运行结果

6. 频繁修改字符串的操作

StringBuffer 应用较多的地方是频繁修改字符串的内容。例如：

```java
public class StringBufferDemo06 {
    public static void main(String[] args) {
        StringBuffer sb = new StringBuffer();            // 声明对象
        sb.append(true);
        for (int i = 0; i < 100; i++) {
            sb.append(i);                                // 比 String 性能高
        }
        System.out.println(sb);
    }
}
```

11.2 Runtime 类

在 Java 中，Runtime 类表示运行时的操作类，是一个封装了 JVM 进程的类，每一个 JVM 都对应着一个 Runtime 类的实例，此实例由 JVM 运行时为其实例化。在 JDK 文档中读者不会发现任何有关 Runtime 类中构造方法的定义，这是因为 Runtime 类本身的构造方法是私有化的。如果想取得一个 Runtime 量的实例，就只能通过以下方式：

```java
Runtime run = Runtime.getRuntime();
```

Runtime 类中提供了一个静态 getRuntime() 方法，此方法可以取得 Runtime 类的实例。Runtime 类的方法如表 11-1 所示。

表11-1 Runtime类中方法

序号	方法定义	描述
1	public static Runtime getRuntime()	取得Runtime类的实例
2	public long freeMemory()	返回Java虚拟机中的空闲内存量
3	public long maxMemory()	返回JVM的最大内存量
4	public void gc()	运行垃圾回收器，释放空间
5	public Process exec(String command) throws IOException	执行本机命令

Runtime 类中方法的使用示例如下：

```java
import java.io.IOException;
public class RuntimeDemo01 {
    public static void main(String[] args) {
        Runtime run = Runtime.getRuntime();
        System.out.println("JVM 最大内存量："+run.maxMemory());
        System.out.println("JVM 空闲内存量："+run.freeMemory());
        String s = "zknu.edu.cn";
        for (int i = 0; i < 1000; i++) {
            s += i;                  // 循环修改 s，产生多个垃圾，会占用内存
        }
        System.out.println(" 循环后 JVM 空闲内存量："+run.freeMemory());
        run.gc();                    // 垃圾回收，释放空间
        System.out.println(" 垃圾回收后 JVM 空闲内存量：
                            "+run.freeMemory());
        Process pro = null;          // 声明一个 Process 对象，接收启动的进程
        try {
            pro = run.exec("calc.exe");       // 调用本机程序
        } catch (IOException e) {
            e.printStackTrace();
        }
        try {
            Thread.sleep(5000);               // 让此线程存活 5 秒
        } catch (InterruptedException e) {
            e.printStackTrace();
        }
        pro.destroy();                        // 结束此线程
    }
}
```

程序运行结果如图 11-6 所示。

图 11-6 程序运行结果

程序调用本机计算器程序，5 秒后自动关闭。

11.3 System 类

System 类是一些与系统相关属性和方法的集合。在 System 类中所有属性都是静态的，要想引用这些属性和方法，直接使用 System 类调用即可。System 类的常用方法如表 11-2 所示。

表11-2 System类的常用方法

序号	方法定义	描述
1	public static void exit(int status)	系统退出
2	public static void gc()	垃圾回收
3	public static long currentTimeMillis()	返回当前时间
4	public static Properties getProperties()	取得当前系统全部属性
5	public static void arraycopy(Object src, int srcPos, Object dest, int destPos, int length)	数组复制操作

System 类的方法相对比较简单，读者可自行练习。

11.4 Math 类

Math 类是数学操作类，其提供了一系列的数学操作方法，包括求绝对值、三角函数等。因为在 Math 类中提供的方法都是静态的，所以直接由类名称调用即可。

下面的示例演示了 Math 类的基本操作。

```java
public class MathDemo {
    public static void main(String[] args) {
        System.out.println("求平方根："+Math.sqrt(9.0));
        System.out.println("求两数的最大值："+Math.max(30, 16));
        System.out.println("求两数的最小值："+Math.min(30, 16));
        System.out.println("2 的 3 次方：: "+Math.pow(2, 3));
        System.out.println("四舍五入："+Math.round(33.6));
    }
}
```

程序运行结果如图 11-7 所示。

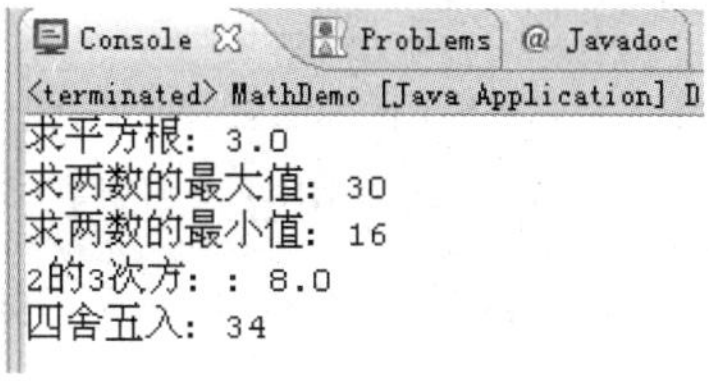

```
Console     Problems  @ Javadoc
<terminated> MathDemo [Java Application] D
求平方根：3.0
求两数的最大值：30
求两数的最小值：16
2的3次方：：8.0
四舍五入：34
```

图 11-7 程序运行结果

11.5 Random 类

Random 类是随机数产生类，可以指定一个随机数的范围，然后任意产生在此范围中的数字。例如，生成 10 个随机数字，且数字不大于 100：

```java
import java.util.Random;
public class RandomDemo {
    public static void main(String[] args) {
        Random rd = new Random();
        for (int i = 0; i < 10; i++) {
            System.out.print(rd.nextInt(100)+"\t");
        }
    }
}
```

程序运行结果如下：（可能的结果）

48 23   80   10   25   22   87   7   78   19

11.6 要点总结

本章通过几个常用类的使用引导读者养成查阅 JDK API 的习惯，这是学好 Java 的关键点之一。

11.7 练习题

1. 填空题

（1）每个 Java 基本类型在 java.lang 包中都有一个相应的包装类，把基本类型数据转换为对象。其中，包装类 Integer 是 _____ 的直接子类。

（2）包装类 Integer 的静态方法可以将字符串类型的数字"123"转换为基本整型变量 n，其实现语句是：_____ Integer.parseInt("123")_____。

（3）使用 java.lang 包中的 _____ StringBuffer/StringBuilder 类创建一个字符串对象，它代表一个字符序列可变的字符串，可以通过相应的方法改变这个字符串对象的字符序列。

（4）StringBuilder 类是 StringBuffer 类的替代类，两者的共同点是都是，可变长度字符串，其中线程安全的类是 _____ StringBuffer _____ 。

（5）使用 Math.random( ) 返回带正号的 double 值，该值大于等于 0.0 且小于 1.0。使用该函数生成 [30,60] 之间的随机整数的语句是 _____ (int)(Math.random()*31)+30 。

2. 选择题

（1）以下选项中，关于 int 和 Integer 的说法错误的是（　　）。

  A. int 是基本数据类型，Integer 是 int 的包装类，是引用数据类型

  B. int 的默认值是 0，Integer 的默认值也是 0

  C. Integer 可以封装了属性和方法提供更多的功能

  D. Integer i=5; 该语句在 JDK1.5 之后可以正确执行，使用了自动拆箱功能

（2）分析如下 Java 代码，该程序编译后的运行结果是（　　）。

```java
public static void main(String[ ] args) {
String str=null;
str.concat("abc");
str.concat("def");
System.out.println(str);
}
```

  A. Null

  B. Abcdef

  C. 编译错误

  D. 运行时出现 NullPointerException 异常

（3）以下关于 String 类的代码执行结果是（　　）。

```java
public class Test2 {
public static void main(String args[]) {
String s1 = new String("bjsxt");
String s2 = new String("bjsxt");
if (s1 == s2)              System.out.println("s1 == s2");
if (s1.equals(s2))         System.out.println("s1.equals(s2)");
}
```

}

  A. s1 == s2　　　B. s1.equals(s2)　　　C. s1 == s2 s1.equals(s2)　　　D. 以上都不对

（4）以下关于 StringBuffer 类的代码执行结果是（　　）。

```
public class TestStringBuffer {
public static void main(String args[]) {
StringBuffer a = new StringBuffer("A");
StringBuffer b = new StringBuffer("B");
mb_operate(a, b);
System.out.println(a + "." + b);
}
static void mb_operate(StringBuffer x, StringBuffer y) {
x.append(y);
y = x;
}
}
```

  A. A.B　　　B. A.A　　　C. AB.AB　　　D. AB.B

第 12 章
Java 项目开发

本章将通过构建两个完整实例的方式介绍使用 Java 语言进行程序设计的完整过程。为了突出重点，前面的章节都是通过一些程序片断来说明各个知识点。学习完 Java 的基础知识之后，最后要掌握的是如何将分散学习的知识综合起来构建一个有现实意义的程序。那么是不是构建程序的过程就是程序实现的过程，即通常所说的编码呢？答案是否定的。一个有实际工作经验的程序员会告诉你编码并不是程序设计的全部。即使构建一个简单的程序，也要经过需求、分析设计、实现和测试的过程。

12.1 软件开发过程

软件开发过程即软件设计思路和方法的一般过程，包括对软件进行需求分析、设计软件的功能与实现的算法和方法、软件的总体结构设计和模块设计、编码和调试、程序联调和测试以及编写、提交程序等一系列操作，以满足客户的需求并解决客户的问题。如果有更高的需求，还需要对软件进行维护、升级处理及报废处理。

12.1.1 需求

在软件行业有一句话：没有需求就没有软件。可见需求是构建软件的出发点，没有人使用的软件或程序是没有现实意义的。即便是简单的程序也有需求，程序所提供的功能要与需求相吻合。程序设计之初明确需求是十分重要的，对于比较大的商业项目，在需求阶段一般通过需求调研获取客户的需求并以用例、需求规格说明书等形式整理记录需求，再配合界面原形确定需求。

求的重要性不容忽视。因为它是程序设计的原点，所以整个程序过程都要围绕这个原点进行。

12.1.2 分析设计

实际上，在软件工程中分析和设计是两个过程。分析的过程是在需求确定的基础上明确软件功能。设计的工程师在分析的基础上设计实现方案用来指导开发实现。通俗地说，分析是明确做什么而设计是明确怎么做。看起来分析的过程与需求过程似乎是重复的，其实它们的区别很明显。需求过程明确的是业务角度的功能，而分析是从（系统）程序的角度考虑功能。对于本章要构建的记事本没有复杂的业务逻辑，功能也比较明确，所以分析的过程可以简化甚至忽略。设计的过程也可以与实现过程同步进行。一般设计过程又分为概要设计和详细设计两个阶段。限于本书篇幅，这里就不详细介绍了。

分析和设计是两个不同的过程。当构建简单程序时，这两个过程可能并不明显，但一定要分清这两个不同过程的工作目标。

12.1.3 实现和测试

一般来说，初学者都关心实现的过程。正如上一小节中提到的，对于简单程序的设计过程和实现过程可以同步进行，所以在下面的两个实例中将重点介绍实现的过程。

程序设计应该是从概要到详细，不断深入、细化的过程，不要过早地陷入程序实现的细节，这样会导致目标不明确很容易偏离需求。这也是构建大型应用程序时，设计过程尤为重要的原因。

实现过程的结束也许会令初学者长出一口气，但程序设计的过程却没有结束，因为构建的程序还未经过测试。虽然编码已经结束，但没有通过测试的程序往往被认为是未完成的程序。限于本书的内容，这里不会介绍测试方法和测试的意义，但要强调一下测试的重要性。

JUnit 是业界非常流行的一种测试框架，有兴趣的读者可以查找相关资料进行专项学习。

12.2 项目实例：记事本工具的开发

12.2.1 需求分析设计

本节要构建的记事本工具参照了 Windows 操作系统中的记事本工具，并在其基础上简化了功能。所以不需要专门撰写用例规约和需求规格说明书，使用功能列表和界面原型表示就可以了。

功能包括以下内容：

（1）保留文件菜单中的新建、打开、保存、另存为和退出功能。

（2）保留编辑菜单中的剪切、复制、粘贴、删除和全选功能。

（3）增加颜色菜单用来设置文本的显示颜色。

（4）保留帮助菜单的关于功能。

（5）保留右键弹出快捷菜单中的剪切、复制、粘贴、删除功能。

界面原型如图 12-1 所示。

图 12-1 界面原型

12.2.2 实现和测试

1. 创建记事本类

首先构建一个名为 Notepad 的类并继承 JFrame 类作为最底层的容器。

```
public class Notepad extends JFrame {
}
```

后面将在此基础上逐步设计实现完成这个简单的记事本。这是一个不断添加代码逐步完善功能的过程，最终完成一个完整的程序。

由于程序中涉及的知识绝大部分在前面章节中已经做过介绍，因此后面将不对程序实现细节做过多的赘述，请读者把学习重心放在程序构建时分析设计的过程。如果遇到程序细节的问题，可参看前面的章节或查阅相关资料。

第 12 章　Java 项目开发

构造一个 Notepad 类时要做三项工作：初始化容器、初始化组件及设置事件监听器。Notepad 类的构造方法如下：

```java
/**
 * 构造方法
 */
public Notepad() {
    // 初始化容器
    initContainer();
    // 初始化组件
    initComponent();
    // 添加事件监听器
    initListener();
}
```

随着一步步的分析设计，将依次定义 initContainer()、initComponent() 和 initListener() 三种方法。

2. 添加组件

在初始化容器之前，首先确定需要哪些组件来组成这个记事本。可以把整个记事本界面分成三大块：菜单栏、文本区和右键菜单。菜单栏和右键菜单如图 12-2 和图 12-3 所示。

图 12-2　菜单栏 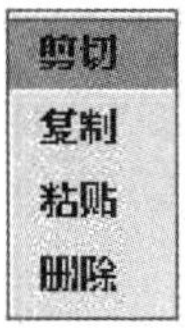　　　　　　　　　　　　　图 12-3　右键菜单

下面在 Notepad 类中定义相关组件类。

```java
/**
 * 内容板
 */
private JPanel contentPane;
/**
 * 菜单栏
 */
private JMenuBar menuBar = new JMenuBar();
/**
 * 文件菜单
 */
private JMenu menuFile = new JMenu();
/**
 * 文件菜单中的新建菜单项
```

```java
     */
    private JMenuItem mItemFileNew = new JMenuItem();
    /**
     * 文件菜单中的打开菜单项
     */
    private JMenuItem mItemFileOpen = new JMenuItem();
    /**
     * 文件菜单中的保存菜单项
     */
    private JMenuItem mItemFileSave = new JMenuItem();
    /**
     * 文件菜单中的另存为菜单项
     */
    private JMenuItem mItemFileSaveAs = new JMenuItem();
    /**
     * 文件菜单中的退出菜单项
     */
    private JMenuItem mItemFileQuit = new JMenuItem();
    /**
     * 编辑菜单
     */
    private JMenu menuEdit = new JMenu();
    /**
     * 编辑菜单中的剪切菜单项
     */
    private JMenuItem mItemEditCut = new JMenuItem();
    /**
     * 编辑菜单中的复制菜单项
     */
    private JMenuItem mItemEditCopy = new JMenuItem();
    /**
     * 编辑菜单中的粘贴菜单项
     */
    private JMenuItem mItemEditPaste = new JMenuItem();
    /**
     * 编辑菜单中的删除菜单项
     */
    private JMenuItem mItemEditDelete = new JMenuItem();
    /**
     * 编辑菜单中的全选菜单项
     */
    private JMenuItem mItemEditSelectAll = new JMenuItem();
    /**
     * 颜色菜单
```

```java
     */
    private JMenu menuColor = new JMenu();
    /**
     * 颜色菜单中的设置颜色菜单项
     */
    private JMenuItem mItemFormatColor = new JMenuItem();
    /**
     * 帮助菜单
     */
    private JMenu menuHelp = new JMenu();
    /**
     * 帮助菜单中的关于菜单项
     */
    private JMenuItem mItemHelpAbout = new JMenuItem();
    /**
     * 右键菜单
     */
    private JPopupMenu popupMenu = new JPopupMenu();
    /**
     * 右键菜单中的剪切菜单项
     */
    private JMenuItem mItemPopupCut = new JMenuItem();
    /**
     * 右键菜单中的复制菜单项
     */
    private JMenuItem mItemPopupCopy = new JMenuItem();
    /**
     * 右键菜单中的粘贴菜单项
     */
    private JMenuItem mItemPopupPaste = new JMenuItem();
    /**
     * 右键菜单中的删除菜单项
     */
    private JMenuItem mItemPopupDelete = new JMenuItem();
    /**
     * 文本编辑区
     */
    private JTextArea textArea = new JTextArea();
    /**
     * 滚动条
     */
    private JScrollPane scroller = new JScrollPane();
    /**
     * 剪贴板
```

```
    */
    private Clipboard cb
= Toolkit.getDefaultToolkit().getSystemClipboard();
```

3. 初始化容器

在确定所需要的组件之后就可以初始化容器了。下面定义 Notepad 类构造方法中调用的 initContainer() 方法。

```
    /**
     * 初始化容器
     */
    private void initContainer() {
        contentPane = (JPanel) this.getContentPane();
        contentPane.setLayout(new BorderLayout());
        contentPane.add(textArea, BorderLayout.CENTER);
        contentPane.add("Center", scroller);
        this.setBounds(100, 100, 500, 500);
        this.setFont(new Font(«宋体», Font.PLAIN, 8));
        this.setTitle(«无标题 - 记事本»);
        this.setJMenuBar(menuBar);
    }
```

4. 初始化组件

初始过容器之后开始初始化相关组件。下面将定义 Notepad 类构造方法中调用的 initComponent() 方法。

```
    /**
     * 初始化组件
     */
    private void initComponent() {
        // 构建菜单栏
        buildMenuBar();
        // 构建右键菜单
        buildPopupMenu();
        // 构建文本区
        buildTextArea();
    }
```

如前面所说，整个记事本界面分成三大块：菜单栏、文本区和右键菜单。所以初始化组件的过程也由构建菜单栏、构建右键菜单和构建文本区三方面组成。

（1）菜单栏

首先构建菜单栏。因为菜单栏又由菜单栏本身和各下拉菜单组成，所以也由 5 个部分构成。

```java
/**
 * 构建菜单栏
 */
private void buildMenuBar() {
    // 菜单栏
    initMenuBar();
    // 文件菜单
    initMenuFile();
    // 编辑菜单
    initMenuEdit();
    // 颜色菜单
    initMenuColor();
    // 帮助菜单
    initMenuHelp();
}
```

下面是这 5 个部分对应的方法定义。

菜单栏如图 12-2 所示。

```java
/**
 * 设置菜单栏
 */
private void initMenuBar() {
    menuBar.add(menuFile);
    menuBar.add(menuEdit);
    menuBar.add(menuColor);
    menuBar.add(menuHelp);
}
```

文件菜单如图 12-4 所示。

图 12-4 文件菜单

```java
/**
 * 设置文件菜单
 */
private void initMenuFile() {
```

```java
        // 设置文件菜单
        menuFile.setText(" 文件 ");
        menuFile.add(mItemFileNew);
        menuFile.add(mItemFileOpen);
        menuFile.add(mItemFileSave);
        menuFile.add(mItemFileSaveAs);
        menuFile.addSeparator();
        menuFile.add(mItemFileQuit);

        // 设置文件菜单中的新建菜单项
        mItemFileNew.setText(" 新建 (N)");
        mItemFileNew.setMnemonic(KeyEvent.VK_N);
        mItemFileNew.setAccelerator(KeyStroke.
getKeyStroke(KeyEvent.VK_N,Event.CTRL_MASK));

        // 设置文件菜单中的打开菜单项
        mItemFileOpen.setText(" 打开 (O)...");
        mItemFileOpen.setMnemonic(KeyEvent.VK_O);
        mItemFileOpen.setAccelerator(KeyStroke.
getKeyStroke(KeyEvent.VK_O,Event.CTRL_MASK));

        // 设置文件菜单中的保存菜单项
        mItemFileSave.setText(" 保存 (S)...");
        mItemFileSave.setMnemonic(KeyEvent.VK_S);
        mItemFileSave.setAccelerator(KeyStroke.
getKeyStroke(KeyEvent.VK_S,Event.CTRL_MASK));

        // 设置文件菜单中的另存为菜单项
        mItemFileSaveAs.setText(" 另存为 (A)...");
        mItemFileSaveAs.setMnemonic(KeyEvent.VK_A);
        // 设置文件菜单中的退出菜单项
        mItemFileQuit.setText(" 退出 ");
    }
```

编辑菜单如图 12-5 所示。

图 12-5 编辑菜单

```java
    /**
     * 设置编辑菜单
     */
```

第 12 章　Java 项目开发

```java
    private void initMenuEdit() {

        // 设置编辑菜单
        menuEdit.setText(" 编辑 ");
        menuEdit.add(mItemEditCut);
        menuEdit.add(mItemEditCopy);
        menuEdit.add(mItemEditPaste);
        menuEdit.add(mItemEditDelete);
        menuEdit.add(mItemEditSelectAll);

        // 设置编辑菜单中的剪切菜单项
        mItemEditCut.setText(" 剪切 (T)");
        mItemEditCut.setMnemonic(KeyEvent.VK_T);
        mItemEditCut.setAccelerator(KeyStroke.
getKeyStroke(KeyEvent.VK_X,Event.CTRL_MASK));
        // 设置编辑菜单中的复制菜单项
        mItemEditCopy.setText(" 复制 (C)");
        mItemEditCopy.setMnemonic(KeyEvent.VK_C);
        mItemEditCopy.setAccelerator(KeyStroke.
getKeyStroke(KeyEvent.VK_C,Event.CTRL_MASK));
        // 设置编辑菜单中的粘贴菜单项
        mItemEditPaste.setText(" 粘贴 (P)");
        mItemEditPaste.setMnemonic(KeyEvent.VK_P);
        mItemEditPaste.setAccelerator(KeyStroke.
getKeyStroke(KeyEvent.VK_V,Event.CTRL_MASK));
        // 设置编辑菜单中的删除菜单项
        mItemEditDelete.setText(" 删除 (L)");
        mItemEditDelete.setMnemonic(KeyEvent.VK_D);
        mItemEditDelete.setAccelerator(KeyStroke.
getKeyStroke(KeyEvent.VK_D,Event.CTRL_MASK));
        // 设置编辑菜单中的全选菜单项
        mItemEditSelectAll.setText(" 全选 (A)");
        mItemEditSelectAll.setMnemonic(KeyEvent.VK_A);
        mItemEditSelectAll.setAccelerator(KeyStroke.
getKeyStroke(KeyEvent.VK_A,Event.CTRL_MASK));
    }
```

颜色菜单如图 12-6 所示。

图 12-6　颜色菜单

```java
    /**
     * 设置颜色菜单
     */
    private void initMenuColor() {
```

```
    // 设置颜色菜单
    menuColor.setText(" 颜色 ");
    menuColor.add(mItemFormatColor);
    // 设置颜色菜单中的颜色菜单项
    mItemFormatColor.setText(" 设置颜色 (C)...");
    mItemFormatColor.setMnemonic(KeyEvent.VK_C);
}
```

帮助菜单如图 12-7 所示。

图 12-7 帮助菜单

```
/**
 * 设置帮助菜单
 */
private void initMenuHelp() {
    // 设置帮助菜单
    menuHelp.setText(" 帮助 ");
    menuHelp.add(mItemHelpAbout);
    // 设置帮助菜单中的关于菜单项
    mItemHelpAbout.setText(" 关于记事本 (A)");
    mItemHelpAbout.setMnemonic(KeyEvent.VK_A);
}
```

（2） 右键菜单

构建完菜单栏之后开始构建右键菜单，如图 12-3 所示。

```
/**
 * 设置右键菜单
 */
private void buildPopupMenu() {
    // 设置右键菜单
    popupMenu.add(mItemPopupCut);
    popupMenu.add(mItemPopupCopy);
    popupMenu.add(mItemPopupPaste);
    popupMenu.add(mItemPopupDelete);

    // 设置右键菜单中的剪切菜单项
    mItemPopupCut.setText(" 剪切 ");
    // 设置右键菜单中的复制菜单项
    mItemPopupCopy.setText(" 复制 ");
    // 设置右键菜单中的粘贴菜单项
```

```
        mItemPopupPaste.setText(" 粘贴 ");
        // 设置右键菜单中的删除菜单项
        mItemPopupDelete.setText(" 删除 ");
    }
```

（3）文本区

最后构建文本区。

```
    /**
     * 构建文本区
     */
    private void buildTextArea() {
        textArea.setRows(20);
        textArea.setColumns(20);
        textArea.setDoubleBuffered(false);
        textArea.setToolTipText(" 这是一个简单的记事本 ");
        textArea.setVerifyInputWhenFocusTarget(true);
        textArea.setText("");
        textArea.add(popupMenu);
        scroller.getViewport().add(textArea);
    }
```

到此为止，初始化组件的方法结束。

5．添加事件监听器

虽然记事本的容器和相关组件都已经初始化完毕，但是要使记事本完成需求中规定的功能就必须添加事件监听器。不仅要为各菜单项和文本区添加监听器，而且还要为容器添加监听器。下面定义添加事件监听器的 initListener() 方法。

```
    /**
     * 添加事件监听器
     */
    private void initListener() {
        ActListener actlistener = new ActListener();
        mItemFileNew.addActionListener(actlistener);
        mItemFileOpen.addActionListener(actlistener);
        mItemFileSave.addActionListener(actlistener);
        mItemFileSaveAs.addActionListener(actlistener);
        mItemFileQuit.addActionListener(actlistener);
        mItemEditCut.addActionListener(actlistener);
        mItemEditCopy.addActionListener(actlistener);
        mItemEditPaste.addActionListener(actlistener);
        mItemEditDelete.addActionListener(actlistener)
```

```
            mItemEditSelectAll.addActionListener(actlistener);
            mItemFormatColor.addActionListener(actlistener);
            mItemHelpAbout.addActionListener(actlistener);
            mItemPopupCut.addActionListener(actlistener);
            mItemPopupCopy.addActionListener(actlistener);
            mItemPopupPaste.addActionListener(actlistener);
            mItemPopupDelete.addActionListener(actlistener);
            textArea.addMouseListener(new MouseListener());
            this.addWindowListener(new WindowListener());
        }
```

6. 构建事件监听器

添加了监听器之后自然要定义这些监听器以便处理各种事件。在添加监听器时，一共使用了三个监听器：ActListener、MouseListener 和 WindowListener。其中，ActListener 用来监听并处理所有菜单项为事件源的事件；MouseListener 处理右键单击事件；WindowListener 用来处理容器关闭触发的事件。

（1）ActListener

ActListener 实现了 ActionListener 接口，其定义如下：

```
            class ActListener implements ActionListener {
            public void actionPerformed(ActionEvent e) {
                    if (e.getSource() == mItemFileNew) {
                        fileNew();
                    } else if (e.getSource() == mItemFileOpen) {
                        fileOpen();
                    } else if (e.getSource() == mItemFileSave) {
                        fileSave();
                    } else if (e.getSource() == mItemFileSaveAs) {
                        fileSaveAs();
                    } else if (e.getSource() == mItemFileQuit) {
                        fileQuit();
                    } else if (e.getSource() == mItemEditCut) {
                        fileCut();
                    } else if (e.getSource() == mItemEditCopy) {
                        fileCopy();
                    } else if (e.getSource() == mItemEditPaste) {
                        filePaste();
                    } else if (e.getSource() == mItemEditDelete) {
                        fileDel();
                    } else if (e.getSource() == mItemPopupCut) {
                        fileCut();
                    } else if (e.getSource() == mItemPopupCopy) {
```

```java
                fileCopy();
            } else if (e.getSource() == mItemPopupPaste) {
                filePaste();
            } else if (e.getSource() == mItemPopupDelete) {
                fileDel();
            } else if (e.getSource() == mItemEditSelectAll) {
                selectAll();
            } else if (e.getSource() == mItemFormatColor) {
                chooseColor();
            } else if (e.getSource() == mItemHelpAbout) {
                about();
            }

        }
    }
```

设计时采用菜单项与处理方法一一对应的方式。由于考虑到处理文件时会用到文件名和文件是否保存的标志，因此所以在类中添加这两个成员变量：

```java
    /**
     * 文件名
     */
    String fileName = null;
    /**
     * 文件是否保存
     */
    boolean isSaved = false;
```

菜单对应的处理方法定义如下：

```java
    /**
     * 新建
     */
    private void fileNew() {
        if (isSaved) {
            this.textArea.setText("");
            this.textArea.setFocusable(true);
            this.setTitle(" 无标题 - 记事本 ");
        } else {
            int result = JOptionPane.showConfirmDialog(this,
" 想保存文件么？ ", " 记事本 ",                JOptionPane.YES_NO_CANCEL_OPTION);
            if (JOptionPane.OK_OPTION == result) {
                fileSaveAs();
            } else if (JOptionPane.NO_OPTION == result) {
                this.textArea.setText("");
```

```java
                    this.textArea.setFocusable(true);
                    this.setTitle("无标题 - 记事本");
                } else {
                }
            }
        }
        /**
         * 打开
         */
        private void fileOpen() {
            String openFileName = "";
            JFileChooser jFileChooser = new JFileChooser();
            if (isSaved) {
                try {
                    if (JFileChooser.APPROVE_OPTION == jFileChooser
                            .showOpenDialog(this)) {
                        openFileName = jFileChooser.getSelectedFile().getPath();
                        File file = new File(openFileName);
                        int flength = (int) file.length();
                        int num = 0;
                        FileReader fReader = new FileReader(file);
                        char[] data = new char[flength];
                        while (fReader.ready()) {
                            num += fReader.read(data, num, flength - num);
                        }
                        fReader.close();
                        textArea.setText(new String(data, 0, num));
                        fileName = openFileName;
                        this.setTitle(fileName.substring(fileName.lastIndexOf("\\") + 1));
                        this.repaint();
                        isSaved = false;
                    }
                } catch (Exception e) {
                    JOptionPane.showMessageDialog(this, "打开文件时出错！", "错误", JOptionPane.ERROR_MESSAGE);
                }
            } else {
                int result = JOptionPane.showConfirmDialog(this, "想保存文件么？", "记事本",JOptionPane.YES_NO_CANCEL_OPTION);
                if (JOptionPane.OK_OPTION == result) {
                    fileSave();
```

```java
                    fileOpen();
                } else if (JOptionPane.NO_OPTION == result) {
                    isSaved = true;
                    fileOpen();
                } else {
                }
            }
        }
        /**
         * 保存
         */
        private void fileSave() {
            if (fileName == null) {
                fileSaveAs();
            } else {
                if (!isSaved) {
                    if (fileName.length() != 0) {
                        try {
                            File saveFile = new File(fileName);
                            FileWriter fw = new FileWriter(saveFile);
                            fw.write(textArea.getText());
                            fw.close();
                            isSaved = true;
                            this.setTitle(fileName.substring(fileName
                                    .lastIndexOf("\\") + 1));
                            this.repaint();
                        } catch (Exception e) {
                            JOptionPane.showMessageDialog(this, "保存文件时出错！", "错误",JOptionPane.ERROR_MESSAGE);
                        }
                    } else {
                        fileSaveAs();
                    }
                }
            }
        }
        /**
         * 另存为
         */
        private void fileSaveAs() {
```

```java
                JFileChooser jFileChooser = new JFileChooser();
                if (JFileChooser.APPROVE_OPTION == jFileChooser.
    showSaveDialog(this)) {
                    fileName = jFileChooser.getSelectedFile().getPath();
                    fileSave();
                }
        }
        /**
         * 退出
         *
         */
        private void fileQuit() {
            if (!isSaved) {
                int result = JOptionPane.showConfirmDialog(this,
    "想保存文件么？", "记事本",
                        JOptionPane.YES_NO_CANCEL_OPTION);
                if (JOptionPane.OK_OPTION == result) {
                    fileSave();
                } else if (JOptionPane.NO_OPTION == result) {
                    System.exit(0);
                } else {
                }
            } else {
                System.exit(0);
            }
        }
        /**
         * 剪切
         */
        private void fileCut() {
            try {
                String str = this.textArea.getSelectedText();
                if (str.length() != 0) {
                    StringSelection s = new StringSelection(str);
                    cb.setContents(s, s);
                    this.textArea.replaceRange("", this.textArea
                            .getSelectionStart(),this.textArea.
    getSelectionEnd());
                    isSaved = false;
                }
            } catch (Exception ex) {
    JOptionPane.showMessageDialog(this, "剪切时出错！", "错误
    ",JOptionPane.ERROR_MESSAGE);
```

```java
        }
    }

    /**
     * 复制
     */
    private void fileCopy() {
        try {
            String str = this.textArea.getSelectedText();
            if (str.length() != 0) {
                StringSelection s = new StringSelection(str);
                cb.setContents(s, s);
            }
        } catch (Exception ex) {
            JOptionPane.showMessageDialog(this, "复制时出错！", "错误",JOptionPane.ERROR_MESSAGE);
        }
    }
    /**
     * 粘贴
     */
    private void filePaste() {
        try {
            Transferable tr = cb.getContents(this);
            if (tr != null) {
                String s = (String) tr.getTransferData(DataFlavor.stringFlavor);
                if (s != null) {
                    textArea.replaceRange(s,textArea.getSelectionStart(),textArea.getSelectionEnd());
                }
                isSaved = false;
            }
        } catch (Exception err) {
            JOptionPane.showMessageDialog(this, "粘贴时出错！", "错误",JOptionPane.ERROR_MESSAGE);
        }
    }
    /**
     * 删除
     */
    private void fileDel() {
        textArea.replaceRange("", textArea.getSelectionStart(), textArea.getSelectionEnd());
        isSaved = false;
    }
```

```java
/**
 * 全选
 */
private void selectAll() {
    textArea.setSelectionStart(0);
    textArea.setSelectionEnd
(this.textArea.getText().length());
}
/**
 * 设置颜色
 */
private void chooseColor() {
    Color bcolor = textArea.getForeground();
    JColorChooser jColor = new JColorChooser();
    jColor.setColor(bcolor);
    textArea.setForeground(JColorChooser.showDialog
(textArea, "选择颜色",bcolor));
}
/**
 * 关于
 */
private void about() {
    JOptionPane.showMessageDialog(this, "这是一个简单的记事本！
", "记事本",    JOptionPane.INFORMATION_MESSAGE);
}
```

（2）MouseListener

MouseListener 继承了 MouseAdapter 类并覆盖 mouseReleased 方法，其定义如下：

```java
class MouseListener extends MouseAdapter {
    @Override
    public void mouseReleased(MouseEvent e) {
        if (e.isPopupTrigger()) {
            popupMenu.show((JTextArea) e.getSource(),
e.getX(), e.getY());
        }
    }
}
```

（3）WindowListener

WindowListener 继承了 WindowAdapter 类并覆盖 windowClosing 方法，其定义如下：

```java
class WindowListener extends WindowAdapter {
    @Override
```

```
        public void windowClosing(WindowEvent arg0) {
                fileQuit();
        }
    }
```

7. 添加显示方法

为 Notepad 类添加一种显示方法，是目前为止除构造方法外 Notepad 类第一个用 public 修饰词修饰的方法。因为只有这种方法供外界调用，其他方法对外界来说都是透明的。显示方法 showNotepad() 定义如下：

```
/**
 * 显示记事本
 */
public void showNotepad() {
      this.setVisible(true);
      this.validate();
}
```

至此，一个简单的记事本工具就构建完成了。

8. 运行测试

添加 main 方法以便运行测试程序。

```
public static void main(String args[]) {
      new Notepad().showNotepad();
}
```

12.3 项目实例：网络通信工具的开发

当前是信息网络得到飞速发展的时代，尤其是计算机与通信技术的发展和结合，深深影响着人们的生活、学习和工作方式。网络聊天工具成为人们日常交流的一种重要工具，其成本低、通信速度快、方便信息交流和资料传递，如微信、QQ 等。本节将进行小型网络通信实例开发。

12.3.1 需求分析设计

本节要构建的网络通信工具功能很简单，包括服务器端和客户端。服务器端启动之后监听端口等待客户端连接。客户端启动之后自动请求连接服务器端，连接建立之后可以实现服务器端和客户端的通信，输入特殊信息后断开连接。为了简化功能，服务器端不要求支持多线程。

界面原型如图 12-8 和图 12-9 所示。

图 12-8 服务器端

图 12-9 客户端

12.3.2 实现和测试

1. 创建服务器端类

构建一个名为 Server 的类并继承 JFrame 类作为最底层的容器。

```
public class Server extends JFrame {
}
```

对于服务器端类构造方法主要任务是初始化组件。Server 的构造方法如下：

```
public Server(String title) {
    super(title);
    initialize();
}
```

2. 添加服务器端组件并初始化

在定义 initialize 方法之前要先确定需要的组件，可参考界面原型。

```
private JTextField textField;
private JTextArea textArea;
```

下面定义 initialize 方法完成组件初始化。

```
private void initialize() {
    textField = new JTextField();
    textField.setEnabled(false);
    textField.addActionListener(new ActionListener() {
        public void actionPerformed(ActionEvent event) {
            sendMessage(event.getActionCommand());
        }
    });
    textArea = new JTextArea();
    Container container = getContentPane();
    container.add(textField, BorderLayout.NORTH);
    container.add(new JScrollPane(textArea), BorderLayout.CENTER);
    setDefaultCloseOperation(JFrame.EXIT_ON_CLOSE);
    setSize(300, 150);
```

```
        setVisible(true);
}
```

3. 添加服务器端运行方法

在完成对组件的初始化之后,开始添加运行服务器端的方法。

```
/**
 * 运行服务器端
 *
 */
public void run() {
    try {
            server = new ServerSocket(PORT);
            while (true) {
                    textArea.setText("等待客户端连接......\n");
                    init();
                    service();
                    disconnect();
                    textArea.append("\n客户端终止连接");
                    COUNTER++;
            }
    } catch (EOFException eofException) {
            System.out.println("客户端终止连接");
    } catch (IOException ioException) {
            System.out.println("I/O异常");
    }
}
```

因为在运行方法中需要使用端口号、计数器和 ServerSocket 的引用,所以下面在类中添加两个静态成员变量分别记录端口号和计数器,添加 ServerSocket 的引用作为成员变量。

```
    private static int COUNTER = 1;
    private static int PORT = 9000;
    private ServerSocket server;
```

因为服务器端运行的过程主要包括等待客户端建立连接及建立连接之后的初始化、接收客户端的消息、发送消息到客户端和断开连接,所以可以设计 init 方法、service 方法、sendMessage 方法和 disconnect 方法对应这些过程。

```
/**
 * 初始化过程
 *
 * @throws IOException
 */
```

```java
    private void init() throws IOException {
        clientConnection = server.accept();
        textArea.append("第 " + COUNTER + " 个连接来自 " +
clientConnection.getInetAddress().getHostName());
        outPutStream = new ObjectOutputStream(clientConnection
                    .getOutputStream());
        outPutStream.flush();
        inPutStream = new ObjectInputStream
(clientConnection.getInputStream());
    }
    /**
     * 处理接收信息
     *
     * @throws IOException
     */
    private void service() throws IOException {
        String message = "服务器端消息：建立连接成功！ ";
        outPutStream.writeObject(message);
        outPutStream.flush();
        textField.setEnabled(true);
        do {
            try {
                message = (String) inPutStream.readObject();
                textArea.append("\n" + message);
                textArea.setCaretPosition(textArea.getText()
.length());
            } catch (ClassNotFoundException e) {
                textArea.append("\n 接收消息时出错 ");
            }
        } while (!message.equals(" 客户端消息：中止通话 "));
    }
    /**
     * 终止连接
     *
     * @throws IOException
     */
    private void disconnect() throws IOException {
        textField.setEnabled(false);
        outPutStream.close();
        inPutStream.close();
        clientConnection.close();
    }
    /**
     * 发送消息
```

```
 *
 * @param message
 */
private void sendMessage(String message) {
    try {
        outPutStream.writeObject("服务器端消息: " + message);
        outPutStream.flush();
        textArea.append("\n 服务器端消息: " + message);
    } catch (IOException ioException) {
        textArea.append("\n 发送消息时出错");
    }
}
```

在 init 方法中需要定义下列成员变量:

```
private ObjectOutputStream outPutStream;
private ObjectInputStream inPutStream;
private Socket clientConnection;
```

至此,服务器端类 Server 构建完成,下面开始构建客户端类。因为客户端运行的过程与服务器端运行的过程基本相同,所以构建客户端类的设计思想与服务器端基本相同。

服务器端运行的过程主要包括等待客户端建立连接及建立连接之后的初始化、接收客户端的消息、发送消息到客户端和断开连接。客户端运行的过程主要包括与服务器端建立连接并初始化、接收客户端的消息、发送消息到客户端和断开连接。

不同之处在于:客户端运行时主动连接服务器端,而服务器端运行时监听端口、等待客户端连接。因此,下面只给出客户端建立连接过程的实现。

4. 客户端建立连接方法

```
/**
 * 与服务器端建立连接
 *
 * @throws IOException
 */
private void connect() throws IOException {
    client = new Socket(InetAddress.getByName(serverIP), PORT);
    textArea.append(" 连接 " + client.getInetAddress().getHostName());
    outPutStream = new ObjectOutputStream(client.getOutputStream());
    outPutStream.flush();
    inPutStream = new ObjectInputStream(client.getInputStream());
```

}

至此，一个简单的网络通信工具就构建完成了。

5. 运行测试

最后分别为 Server 类和 Client 类添加 main 方法以便运行测试程序。

```
public static void main(String args[]) {
    Server server = new Server("服务器端");
    server.run();
}
public static void main(String args[]) {
    Client client = new Client("客户端");
    client.run();
}
```

12.4 项目实例：在线相册的开发

随着网络的发展，现在网上出现了越来越多以用户个人为主要服务对象的应用程序，如个人博客、个人相册等。在线成功申请后，就可以将自己的文章发表到博客与其他用户进行交流，也可以将自己喜欢的照片上传到相册与大家分享。本章将对在线相册的实现进行详细介绍。

12.4.1 需求分析设计

在线相册主要实现用户照片和相册的管理功能，用户登录系统后，可以新建相应类型的相册，可以给已有的相册上传照片，也可以查看相册中的照片或删除照片。如果用户觉得某个相册中的照片有些过时或不喜欢了，也可以将整个相册删除。

下面对整个工作流程做一个详细的介绍。

- 用户打开网页，输入正确的用户名和密码，登录系统进入用户个人相册首页，包括我的相册、新建相册、上传照片、删除相册和退出系统操作。
- 用户可以通过我的相册链接查看已有相册操作（查看相应相册中的照片，并可以单击删除链接删除照片，单击图片名称查看原始图片）。
- 新建相册（选择新建相册的类型，给自己的相册命名并提交建立相册）。
- 用户可以通过上传照片链接进入上传照片页面（选择将照片传到哪个相册，选择上传照片路径）。
- 用户可以通过删除相册链接进入删除相册页面（本页列出了所有用户已有相册及其相册信息，用户可以选择删除）。

- 退出系统（退出相册）。

如图 12-10 所示为系统的架构。

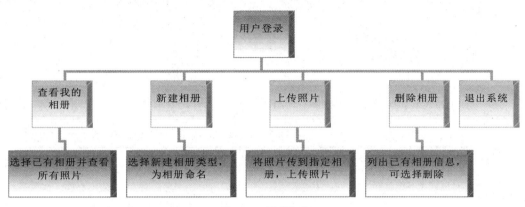

图 12-10 系统架构

12.4.2 数据库设计

数据库设计是整个系统设计中最重要也是最难的一部分。数据库设计是否合理直接关系到系统的实现和系统的灵活性。合理的数据库设计将使系统的编写更加容易，并且更容易扩展新的功能而不需要重新设计数据库。

1. MySQL 存取图片

在介绍具体的数据库实现之前，首先应该了解一下怎么将图片存入 MySQL 数据库，在本系统中，用户上传的照片都将被存入数据库中。

在 MySQL 数据库中保存图片的类型为 blob。因为用户上传的图片可能比较大，所以本节介绍的应用中将使用 mediumblob 类型。例如：

```
photo mediumblob not null,
```

数据库表中的 photo 字段将用于存储 mediumblob 类型的数据。在将图片存入数据库时，首先要使用 I/O 流读入图片文件，比如：

```
// 读取照片
File f=new File(path);
int i=(int)f.length();
byte[] bb=new byte[i];
fis=new FileInputStream(f);
fis.read(bb);
```

然后将获得的 byte 数组写入数据库中。

2. 数据库建表

下面是用于创建数据库的 SQL 语句：

```
/* 删除已有的 photos 数据库 */
drop database photos;
/* 创建 photos 数据库 */
create database photos;
/* 使用 photos 数据库 */
use photos;
```

接下来开始建立数据库表，表 12-1 列出了用户表的相关信息。

表12-1 用户表

字段名	字段类型	字段大小	NULL	默认值
user_name	varchar	20	not null	无
user_pwd	varchar	20	not null	无
UA_id	varchar	10	not null	无

用户表（users）用于存储用户的基本信息，主键为 UA_id 字段，建立主键：

```
constraint user_pk primary key (UA_id)
```

表 12-2 列出了相册类型表的相关信息。

表12-2 相册类型表

字段名	字段类型	字段大小	NULL	默认值
album_type_id	varchar	5	not null	无
album_type_name	varchar	20	not null	无

相册类型表（album_type）用于存储相册类型信息，主键为 album_type_id 字段，建立主键：

```
constraint album_type_pk primary key (album_type_id)
```

表 12-3 列出了用户与用户相册关系表的相关信息。

表12-3 用户与用户相册关系表

字段名	字段类型	字段大小	NULL	默认值
UA_id	varchar	10	not null	无
album_id	varchar	10	not null	无
album_name	varchar	20	not null	无
album_type_id	varchar	5	not null	无
new_time	varchar	20	not null	sysdate()

用户与用户相册关系表（user_album）用于存储用户与用户相册关系信息，主键为 album_id 字段，建立主键及和其他表的关系（外键）：

```
    constraint user_album_pk primary key (album_id),
    constraint user_album_fk foreign key (UA_id) references users(UA_id),
    constraint album_type_fk foreign key (album_type_id) references album_
type(album_type_id)
```

表 12-4 列出了照片表的相关信息。

表12-4 照片表

字段名	字段类型	字段大小	NULL	默认值
photo_id	varchar	10	not null	无
album_id	varchar	10	not null	无
photo_name	varchar	20	not null	无
photo	mediumblob		not null	无

照片表（photos）用于存储用户照片信息，主键为 photo_id 字段，建立主键及和其他表的关系（外键）：

```
    constraint photo_pk primary key (photo_id),
    constraint photo_fk foreign key (album_id) references user_
album(album_id)
```

下面是用于创建数据库及其数据库表的完整的 SQL 语句，并向其中输入了一些测试数据。

```
/* 删除已有的 photos 数据库 */
drop database photos;
/* 创建 photos 数据库 */
create database photos;
/* 使用 photos 数据库 */
use photos;
/* 创建用户表 */
drop table users;
create table users
(
    user_name varchar(20) not null,
    user_pwd varchar(20) not null,
    UA_id varchar(10) not null,
    constraint user_pk primary key (UA_id)
);
/* 创建相册类型表 */
drop table album_type;
create table album_type
(
    album_type_id varchar(5) not null,
    album_type_name varchar(20) not null,
    constraint album_type_pk primary key (album_type_id)
```

```sql
);
/* 创建用户相册关系表 */
drop table user_album;
create table user_album
(
    UA_id varchar(10) not null,
    album_id varchar(10) not null,
    album_name varchar(20) not null,
    album_type_id varchar(5) not null,
    new_time varchar(20) not null,
    constraint user_album_pk primary key (album_id),
    constraint user_album_fk foreign key (UA_id) references users(UA_id),
    constraint album_type_fk foreign key (album_type_id) references album_type(album_type_id)
);
/* 创建照片表 */
drop table photos;
create table photos
(
    photo_id varchar(10) not null,
    album_id varchar(10) not null,
    photo_name varchar(20) not null,
    photo mediumblob not null,
    constraint photo_pk primary key (photo_id),
    constraint photo_fk foreign key (album_id) references user_album(album_id)
);
/* 插入用户名和密码及用户相册 */
insert into users values('root','123456','54632345');
/* 插入相册类型 */
insert into album_type values('10001','我的家庭');
insert into album_type values('10002','个性生活');
insert into album_type values('10003','朋友圈');
insert into album_type values('10004','旅游见闻');
/* 用户相册信息 */
insert into user_album values('54632345','53468132','我的相册1','10001',sysdate());
```

到此为止，数据库建立完成。

12.4.3 开发数据库 JavaBean

对数据库的操作在整个应用程序中是相对独立的部分。因为在系统的其他任何一个模块中都可能用到不同的数据库操作，所以在实际开发中往往是将对数据库操作的部分提取出来，作为开发的第一步。这样便于以后编码过程中的按模块分工，使得整个开发过程更容易管理和进行。下面就来讲解本例的数据库 JavaBean 的开发。

通过前面的需求分析，已经知道了本系统所要完成的功能和操作。本例中的数据访问对象是一个 JavaBean 类，封装了对数据库的所有操作，包括连接数据库及查询、更新、插入和删除数据库数据。

首先，在 JCreator 中新建一个名称为 AlbumDAOBean 的 Java 类，它是公有的（public）并位于 com.album 包中。输入以下代码：

```java
package com.album;
import java.io.*;
import java.sql.*;
import java.sql.Connection;
import javax.sql.*;
import java.util.*;
import javax.naming.*;
import com.javabean.AlbumPhoto;
import com.javabean.AlbumInfo;
import com.javabean.Photo;
public class AlbumDAOBean {
    ///onlinealbumJNDI 是 JNDI 名称
    private static final String JNDI_STR=
                                    "java:comp/env/onlinealbumJNDI";
    private static Context ctx;
    private static Connection con;
    private static ResultSet rs;
    private static PreparedStatement userLogInPS;
    private static PreparedStatement photoPS;
    private static PreparedStatement insertPhotoPS;
    private static PreparedStatement UAIdPS;
    private static PreparedStatement newAlbumPS;
    private static PreparedStatement albumInfoPS;
    private static PreparedStatement delAlbumPS;
    private static PreparedStatement delAlbumDetailPS;
    private static PreparedStatement delPhotoPS;
    private static PreparedStatement albumTypeInfoPS;
    private static PreparedStatement albumPhotoInfoPS;
    static {// 预编译 SQL
        try {
            // 取得数据库连接
            con=AlbumDAOBean.getConnection();
            // 用户登录
            userLogInPS=con.prepareStatement(
```

```java
            "SELECT user_pwd FROM users WHERE user_name=?");
        //从数据库提取照片
        photoPS=con.prepareStatement(
            "SELECT photo FROM photos WHERE photo_id=?");
        //上传照片
        insertPhotoPS=con.prepareStatement(
            "INSERTINTO photos(album_id,photo_id,photo_name,photo) VALU"
            +"ES(?,?,?,?)");
        //用户与相册关系
        UAIdPS=con.prepareStatement(
            "SELECT UA_id FROM users WHERE user_name=?");
        //新建相册
        newAlbumPS=con.prepareStatement(
        "INSERT INTuser_album(UA_id,album_id,album_name,album_type"
            +"_id,new_time) VALUES(?,?,?,?,sysdate())");
        //相册详细信息
        albumInfoPS=con.prepareStatement(
        SELECTua.album_id,ua.album_name,ua.new_time,(SELECT at.alb"
            +"um_type_name FROM album_type at WHERE at.album_type_id=ua."
            +"album_type_id),(SELECT COUNT(p.photo_id) FROM photos p WHE"
            +"RE p.album_id=ua.album_id),(SELECT at1.album_type_id FROM "
            +"album_type at1 WHERE at1.album_type_id=ua.album_type_id) F"
            +"ROM user_album ua,users u WHERE u.user_name=?");
        //删除所有用户相册照片
        delAlbumDetailPS=con.prepareStatement(
            "DELETE FROM photos WHERE album_id=?");
        //删除相册
        delAlbumPS=con.prepareStatement(
            "DELETE FROM user_album WHERE album_id=?");
        //删除照片
        delPhotoPS=con.prepareStatement(
            "DELETE FROM photos WHERE photo_id=?");
        //相册类型信息
        albumTypeInfoPS=con.prepareStatement(
            "SELECT album_type_id,album_type_name FROM album_type");
```

```
                    //相册照片集
                    albumPhotoInfoPS=con.prepareStatement(
                        "SELECT photo_id,photo_name,album_id FROM photos WHERE"
                        +" album_id=?");
            } catch (SQLException se) {
                se.printStackTrace();
            }
        }
    }
```

上面的代码只是这个 JavaBean 类的框架,在这里声明了需要用到的 PreparedStatement 对象和预编译的 SQL 语句。详细功能说明可以在代码注释中找到,这里不再一一列出。

有了框架之后就可以向其中添加具体的功能代码,即功能方法。从用户的登录验证开始,下面是用于用户登录验证的方法。这个方法是公有的(public)的,并且是静态的(static)方法,因为在非静态语句块中不能访问静态的成员,也就是前面声明的预编译的 SQL 语句。userLogIn 方法的代码如下:

```
//方法名:userLogIn
//功能介绍:用户登录验证
//参数说明:userName:用户名 passWord:密码
//返 回 值:false/true
//异常:java.sql.SQLException
public static boolean userLogIn(String userName,String passWord) {
    boolean flag=false;
    String pwd=null;
    try {
        userLogInPS.setString(1,userName);
        rs=userLogInPS.executeQuery();
        while(rs.next()) {
            pwd=rs.getString(1).trim();
        }
        //判断密码是否正确
        if(passWord.equals(pwd)) {
            flag=true;
        }
    } catch (SQLException se) {
        se.printStackTrace();
        flag=false;
    } finally {// 关闭结果集
        close(rs);
    }
    return flag;
```

}

该方法通过用户名和密码来验证用户登录（先查找数据库中相应用户名的密码，然后比较此密码与用户输入的密码是否相同），如果用户输入的信息正确，就返回 true，否则返回 false。

用户成功登录系统后，可以对自己的相册做相应的操作。getAlbumInfo 方法的代码如下：

```java
// 方法名：getAlbumInfo
// 功能介绍：取得用户相册信息
// 参数说明：userName：用户名
// 返 回 值：java.util.Vector
// 异常 :java.sql.SQLException
//java.io.UnsupportedEncodingException
public static Vector getAlbumInfo(String userName) {
    Vector vec=new Vector();
    try {
        albumInfoPS.setString(1,userName);
        rs=albumInfoPS.executeQuery();
        while(rs.next()) {
            //com.javabean.AlbumInfo
            AlbumInfo ai=new AlbumInfo();
            ai.setAlbumId(rs.getString(1));
            ai.setAlbumName(
                // 将 MySQL 编码 ISO-8859-1 转换为国标2312(gb2312)
                new String(rs.getString(2).getBytes(
                    "ISO-8859-1"),"gb2312"));
            ai.setNewTime(rs.getString(3));
            ai.setAlbumType(
                new String(rs.getString(4).getBytes(
                    "ISO-8859-1"),"gb2312"));
            ai.setPhotoCount(rs.getInt(5));
            ai.setAlbumTypeId(rs.getString(6));
            vec.add(ai);
            ai=null;
        }
    } catch (SQLException se) {
        se.printStackTrace();
    } catch (UnsupportedEncodingException uee) {
        uee.printStackTrace();
    } finally {// 关闭结果集
        close(rs);
    }
}
```

```
            return vec;
    }
```

此方法通过用户的用户名来获取用户相册的信息,用户名从 Servlet 的 session 对象中获得(在用户成功登录后存入 session 对象中),以便显示当前登录用户的相册信息。这里将从数据库中提取的信息存入另一个 JavaBean 中,下面是 AlbumInfo.java 的代码:

```
package com.javabean;
import java.io.*;
public class AlbumInfo implements Serializable {
    private String albumId;
    private String albumName;
    private String albumType;
    private String albumTypeId;
    private String newTime;
    private int photoCount;
    public void setAlbumTypeId(String albumTypeId) {
            this.albumTypeId=albumTypeId;
    }
    public String getAlbumTypeId() {
            return (this.albumTypeId);
    }
    public void setAlbumType(String albumType) {
            this.albumType=albumType;
    }
    public String getAlbumType() {
            return (this.albumType);
    }
    public void setAlbumId(String albumId) {
            this.albumId=albumId;
    }
    public String getAlbumId() {
            return (this.albumId);
    }
    public String getAlbumName() {
            return (this.albumName);
    }
    public void setAlbumName(String albumName) {
            this.albumName = albumName;
    }
    public String getNewTime() {
            return (this.newTime);
    }
    public void setNewTime(String newTime) {
```

```java
            this.newTime = newTime;
    }
    public int getPhotoCount() {
            return (this.photoCount);
    }
    public void setPhotoCount(int photoCount) {
            this.photoCount = photoCount;
    }
    public String toString() {
            String sep = System.getProperty("line.separator");
            StringBuffer buffer = new StringBuffer();
            buffer.append(sep);
            buffer.append("albumId = ");
            buffer.append(albumId);
            buffer.append(sep);
            buffer.append("albumName = ");
            buffer.append(albumName);
            buffer.append(sep);
            buffer.append("newTime = ");
            buffer.append(newTime);
            buffer.append(sep);
            buffer.append("photoCount = ");
            buffer.append(photoCount);
            buffer.append(sep);
            return buffer.toString();
    }
}
```

在从数据库中提取数据时，如果有中文，就必须将其转换为中文编码，否则将显示为乱码。下面是 getAlbumTypeInfo 方法的代码，此方法用于获取相册类型信息。

```java
// 方法名：getAlbumTypeInfo
// 功能介绍：取得相册类型信息
// 参数说明：没有参数
// 返回值：java.util.HashMap
// 异常 :java.sql.SQLException
//      java.io.UnsupportedEncodingException
public static HashMap getAlbumTypeInfo() {
    HashMap hm=new HashMap();
    try {
            rs=albumTypeInfoPS.executeQuery();
            while(rs.next()) {
                    hm.put(rs.getString(1),
                    // 将 MySQL 编码 ISO-8859-1 转换为国标 2312(gb2312)
```

```
                    new String(rs.getString(2).getBytes("ISO-8859-
1"),"gb2312"));
            }
        } catch (SQLException se) {
            se.printStackTrace();
        } catch (UnsupportedEncodingException uee) {
            uee.printStackTrace();
        } finally {// 关闭结果集
            close(rs);
        }
        return hm;
    }
```

用户在新建相册时可以选择所建相册的类型，以便进行管理，这里将相册的类型信息存入一个 HashMap 中。下面是 newAlbum 方法的代码，此方法用于新建相册。

```
// 方法名: newAlbum
// 功能介绍: 新建相册
// 参数说明: uaId:用户和相册关系编号  albumId:照片编号
//type_id: 相册类型编号  album_name: 相册名称
// 返回值: false/true
// 异常 :java.sql.SQLException
//       java.io.UnsupportedEncodingException
public static boolean newAlbum(String uaId,String album_id,
                                    String type_id,String album_name) {
    boolean b=false;
    try {
            newAlbumPS.setString(1,uaId);
            newAlbumPS.setString(2,album_id);
            newAlbumPS.setString(3,
            // 将 MySQL 编码 ISO-8859-1 转换为国标 2312(gb2312)
                    new String(album_name.getBytes(),"ISO-8859-1"));
            newAlbumPS.setString(4,type_id);
            newAlbumPS.execute();
            b=true;
    } catch (SQLException se) {
            se.printStackTrace();
            b=false;
    } catch (UnsupportedEncodingException uee) {
            uee.printStackTrace();
        }
        return b;
    }
```

deleteAlbum 方法可以删除用户所建的相册，同时删除当前相册中的所有照片。代码如下：

```java
// 方法名：deleteAlbum
// 功能介绍：删除用户相册
// 参数说明：albumId: 相册编号
// 返回值：false/true
// 异常 :java.sql.SQLException
public static boolean deleteAlbum(String albumId) {
    boolean b=false;
    try {
        delAlbumDetailPS.setString(1,albumId);
        delAlbumDetailPS.execute();
        delAlbumPS.setString(1,albumId);
        delAlbumPS.execute();
        b=true;
    } catch (SQLException se) {
        se.printStackTrace();
        b=false;
    }
    return b;
}
```

如果删除成功，就返回 true，否则返回 false。

当用户上传照片时，insertImage 方法被调用。代码如下：

```java
// 方法名：insertImage
// 功能介绍：上传照片
// 参数说明：albumId: 相册编号 photo_id: 照片编号
//          photo_name: 照片名称 path: 路径
// 返回值：false/true
// 异常 :java.sql.SQLException
//      java.io.IOException
public static boolean insertImage(String album_id,String photo_id,
                                  String photo_name,String path) {
    boolean b=false;
    FileInputStream fis=null;
    try {
        // 读取照片
        File f=new File(path);
        int i=(int)f.length();
        byte[] bb=new byte[i];
        fis=new FileInputStream(f);
        fis.read(bb);
```

```
            insertPhotoPS.setString(1,album_id);
            insertPhotoPS.setString(2,photo_id);
            insertPhotoPS.setString(3,
            // 将MYSQL编码ISO-8859-1转换为国标2312(gb2312)
                    new String(photo_name.getBytes(),"ISO-8859-1"));
            insertPhotoPS.setBytes(4,bb);
            insertPhotoPS.execute();
            b=true;
    } catch (IOException ie) {
            ie.printStackTrace();
            b=false;
    } catch (SQLException se) {
            se.printStackTrace();
            b=false;
    } finally {// 关闭流
            try {
                    if(fis!=null) {
                            fis.close();
                    }
            } catch (IOException ie) {
                    ie.printStackTrace();
            }
    }
    return b;
}
```

如果上传成功,就返回 true,否则返回 false。此方法会从参数中的路径信息中读取用户所要上传的照片,然后存入 byte 数组中,最后将此字节数组写入数据库中。GetAlbumPhotoInfo() 方法用于获取用户相应相册中照片的基本信息,此方法的参数 albumId 为用户相册的编号,通过它可以找到相册中的照片。代码如下:

```
// 方法名:getAlbumPhotoInfo
// 功能介绍:取得相册照片
// 参数说明:albumId:相册编号
// 返回值:java.util.Vector
// 异常:java.sql.SQLException
//      java.io.UnsupportedEncodingException
public static Vector getAlbumPhotoInfo(String albumId) {
    Vector Vec=new Vector();
    try {
            albumPhotoInfoPS.setString(1,albumId);
            rs=albumPhotoInfoPS.executeQuery();
            while(rs.next()) {
```

```java
                //com.javabean.Photo
                Photo photo=new Photo();
                photo.setPhotoId(rs.getString(1));
                photo.setPhotoName(
                    // 将MYSQL编码ISO-8859-1转换为国标2312(gb2312)
                    new String(rs.getString(2).getBytes(
                        "ISO-8859-1"),"gb2312"));
                photo.setAlbumId(rs.getString(3));
                Vec.add(photo);
                photo=null;
            }
    } catch (SQLException se) {
        se.printStackTrace();
    } catch (UnsupportedEncodingException uee) {
        uee.printStackTrace();
    } finally {// 关闭结果集
        close(rs);
    }
    return Vec;
}
```

这里将从数据库中提取的信息存入另一个JavaBean(Photo.java)中。下面是Photo.java的代码，用于存储用户相册照片的信息。

```java
package com.javabean;
public class Photo {
    private String photoId;
    private String photoName;
    private String albumId;
    public void setAlbumId(String albumId) {
        this.albumId=albumId;
    }
    public String getAlbumId() {
        return (this.albumId);
    }
    public void setPhotoId(String photoId) {
        this.photoId=photoId;
    }
    public String getPhotoId() {
        return (this.photoId);
    }
    public String getPhotoName() {
        return (this.photoName);
    }
```

```java
    public void setPhotoName(String photoName) {
        this.photoName = photoName;
    }
    public String toString() {
        String sep = System.getProperty("line.separator");
        StringBuffer buffer = new StringBuffer();
        buffer.append(sep);
        buffer.append("photoId = ");
        buffer.append(photoId);
        buffer.append(sep);
        buffer.append("photoName = ");
        buffer.append(photoName);
        buffer.append(sep);
        buffer.append("albumId = ");
        buffer.append(albumId);
        buffer.append(sep);
        return buffer.toString();
    }
}
```

Photo.java 主要用于存储用户相册中照片的编号、名称和此照片所在相册的编号。不仅要将照片的基本信息从数据库中提取出来，最主要的是将照片在网页上显示给用户，所以还需要将照片本身从数据库中提取出来。getPhoto() 就是实现此功能的方法，代码如下：

```java
//方法名：getPhoto
//功能介绍：取得照片
//参数说明：photoId：照片编号
//返回值：java.util.Vector
//异常：java.sql.SQLException
public static Vector getPhoto(String photoId) {
    Vector Vec=new Vector();
    try {
        photoPS.setString(1,photoId);
        rs=photoPS.executeQuery();
        while(rs.next()) {
                //com.javabean.AlbumPhoto
                AlbumPhoto ap=new AlbumPhoto();
                ap.setPhotoByte(rs.getBytes(1));
                Vec.add(ap);
                ap=null;
        }
    } catch (SQLException se) {
        se.printStackTrace();
    } finally {// 关闭结果集
```

```
            close(rs);
    }
    return Vec;
}
```

以上是将从数据库中提取的信息存入另一个 JavaBean(AlbamPhoto.java) 中，下面是 AlbumPhoto.java 类的代码，用于存储用户相册中照片的字节信息。

```
package com.javabean;
public class AlbumPhoto {

    private byte[] photoByte;
    public byte[] getPhotoByte() {
        return (this.photoByte);
    }
    public void setPhotoByte(byte[] photoByte) {
        this.photoByte = photoByte;
    }
}
```

在将照片显示到网页上时，可以直接将此字节数组写入输出流，然后显示照片。具体实现将在本章的后面进行介绍。

用户不仅可以上传照片，也可以删除相册中已有的照片，deletePhoto 方法就实现了这样的功能。代码如下：

```
// 方法名：deletePhoto
// 功能介绍：删除照片
// 参数说明：albumId：照片编号
// 返回值：false/true
// 异常 :java.sql.SQLException
public static boolean deletePhoto(String photoId) {
    boolean b=false;
    try {
        delPhotoPS.setString(1,photoId);
        delPhotoPS.execute();
        b=true;
    } catch (SQLException se) {
        se.printStackTrace();
        b=false;
    }
    return b;
}
```

以上代码直接通过照片的编号找到照片并删除，如果删除成功，就返回 true，否则返回

false。getUA_id()方法用于取得用户和相册关系编号,以便通过用户找到用户相册。代码如下:

```java
// 方法名: getUA_id
// 功能介绍: 取得用户和相册关系编号
// 参数说明: userName: 用户名
// 返回值: false/true
// 异常 :java.sql.SQLException
public static String getUA_id(String userName) {
    String stemp="";
    try {
        UAIdPS.setString(1,userName);
        rs=UAIdPS.executeQuery();
        while(rs.next()) {
            stemp=rs.getString(1);
        }
    } catch (SQLException se) {
        se.printStackTrace();
    }finally {// 关闭结果集
        close(rs);
    }
    return stemp;
}
```

到这里,数据访问对象的主要功能已经完成。下面的代码是获取数据库连接和关闭数据库资源的方法:

```java
// 方法名: getConnection()
// 功能介绍: 取得数据库连接
// 参数说明: 无
// 返回值: java.sql.Connection
// 异常 :java.sql.SQLException
//      javax.naming.NamingException
public static Connection getConnection() {
    DataSource ds=null;
    try {
        //Tomcat 获取 MySQL 数据库连接
        ctx=new InitialContext();
        ds=(DataSource)ctx.lookup(JNDI_STR);
        con=ds.getConnection();
    } catch (SQLException se) {
        se.printStackTrace();
    } catch (NamingException ne) {
        ne.printStackTrace();
    }
```

```
        return con;
}
// 方法名: close
// 功能介绍: 关闭结果集
// 参数说明: rs: 结果集
// 返回值: void
// 异常 :java.sql.SQLException
private static void close(ResultSet rs) {
    try {
            if(rs!=null) {
                    rs.close();
            }
    } catch (SQLException se) {
            se.printStackTrace();
    }
}
```

这里的 getConnection() 方法很好地体现了代码的重用性，因为不用在每个方法中都写获取数据库连接的代码，所以只要通过调用此方法就可以完成。打开的资源在不需要的情况下一定要关闭，如结果集等，如果不关闭，这些大的数据对象就会占用很多内存空间，使得应用程序的性能大大降低。

12.4.4 实现和测试

在各功能模块中，用户提交的请求都被送到 Servlet 进行处理，可以调用 JavaBean 访问数据库，然后将这些数据组织好发回客户端显示给用户。Servlet 主要通过客户端提交的 action 来判断用户的当前操作，以进行相应地处理。下面是本应用中 Servlet 的框架结构，功能代码以后加入即可。

```
import java.io.*;
import java.util.*;
import javax.servlet.*;
import javax.servlet.http.*;
import com.jspsmart.upload.*;
import com.album.AlbumDAOBean;
public class AlbumServlet extends HttpServlet {
    private ServletConfig config;
    // 初始化 Servlet
    public final void init(ServletConfig config) throws ServletException {
            this.config = config;
    }
    // 处理 POST 请求
    public void doPost(HttpServletRequest request,HttpServletResponse
```

```java
response)                                                              throws 
ServletException,IOException {
                request.setCharacterEncoding("GBK");
                response.setCharacterEncoding("GBK");
                // 获取 PrintWriter 对象，用于向客户端输出信息
                PrintWriter out=response.getWriter();
                // 获取 HttpSession 对象
                HttpSession session=request.getSession();
                // 如果 session 对象为空，就返回
                if(session==null) {
                    return;
                }
                // 设置 session 的有效时间为 10 分钟
                session.setMaxInactiveInterval(60*10);
                String action=request.getParameter("action");
    }
    // 处理 GET 请求
    public void doGet(HttpServletRequest request,HttpServletResponse response) 
throws ServletException,IOException {
                this.doPost(request,response);
    }
    // 跳转
    private void forward(HttpServletRequest request,HttpServletResponse response,String url) 
throws ServletException,IOException {
                RequestDispatcher dispatcher=request.getRequestDispatcher(url);
                dispatcher.forward(request,response);
    }
}
```

String action=request.getParameter("action")用于获取客户端的动作信息，可以判断客户当前的操作是什么，然后调用相应的业务逻辑进行处理以返回给用户正确的结果。

1. 用户身份验证模块

相册首页 index.jsp 的代码如下：

```
<%@ page
 contentType="text/html;charset=GBK"
 errorPage="error.jsp"
```

```
    %>
    <html>
      <head>
      <title></title>
      <style type="text/css">
            .STYLE4 {
                font-size: 13px;
                font-family: " 宋体 ";
                color: #244ac6;
                line-height: 16pt;
            }
      </style>
      <script language="JavaScript">
            function login() {
            var login=document.getElementById("login");
            login.style.display=(login.style.display=="none")? "block":"none";
                var log=document.getElementById("log");
                log.style.display=(log.style.display=="none")? "block":"none";
            }
      </script>
      </head>
      <body>
      <form action="AlbumServlet" method="post" name="myform" id="myform">
      <center><br><br><br><br>
      <table><tr><td>
            <div id="log" class="STYLE4" style="display:block"> 单击图片登录您的相册.....
                      <br></div>
            <img src="image/ge.jpg" onclick="JavaScript:login();"><br>
            <div id="login" class="STYLE4" style="display:none">
            <table><tr>
                <td><font color="#30358D"><b> 用户名 : </td></b><td>
                    <input type="text" name="username" id="username" style="width:120">
                         <br></font></td></tr><tr>
                <td><font color="#30358D"><b> 密   码 : </td></b><td>
                    <input type="password" name="password" id="password" style="width:120">
```

```
                <br></font></td></tr>
        <tr height="10"><td></td></tr>
        <tr align="center"><td>
                <input type="submit" value=" 登录 ">
                <input type="hidden" name="action" value="userLogin">
        </td><td>
                <input type="reset" value=" 重置 ">
        </td></tr></table></div></td></tr></table></center></form></body>
    </html>
```

运行结果如图 12-11 所示。

图 12-11 相册首页代码的运行结果

单击图片，会出现如图 12-12 所示的登录页面。

图 12-12 登录页面

这里会用到 JavaScript 的相关知识，主要是实现此处的特效（单击图片显示登录页面），即用户在打开网页时，用于输入用户名和密码的文本框和按钮都不可见，在用户单击图片后将输入部分的属性设置为可见。关于 JavaScript 的相关知识本书不做任何介绍，有兴趣的读者可以参考 JavaScript 的书籍或其他资料进行学习。

<input type="hidden" name="action" value="userLogin"> 即 action（动作），将 action 属性的值设置为 userLogin。在将 action 属性提交给服务器时，Servlet 会判断其属性值，

如这里的值为 userLogin，Servlet 就会调用验证用户登录的 JavaBean 工作，然后将结果返回给客户端。Servlet 中的代码如下：

```
// 用户登录
        if(action.equals("userLogin")) {
        // 如果用户执行的是登录操作，就执行此段代码
            // 收取用户提交的用户名和密码
            String userName=request.getParameter("username").trim();
            String passWord=request.getParameter("password").trim();
        // 调用 JavaBean 验证用户名和密码
            boolean b=AlbumDAOBean.userLogIn(userName,passWord);
            if(b) {
            // 如果登录成功，就将用户信息存入 session 对象中
                // 并跳转至首页显示
                session.setAttribute("userName",userName);
                Vector vec=AlbumDAOBean.getAlbumInfo(userName);
                String uaId=AlbumDAOBean.getUA_id(userName);
                session.setAttribute("uaId",uaId);
                session.setAttribute("userAlbum",vec);
                this.forward(request,response,"/albumindex.jsp");
            } else {// 否则跳转至错误页
                this.forward(request,response,"/error.jsp?msg=用户名或密码错误！");
            }
        }
```

如果用户登录成功，就会出现如图 12-13 所示的页面。

图 12-13 相册首页

如果登录失败，就会跳转到错误页，提示用户登录失败。error.jsp 的代码如下：

```
<%@ page
```

```jsp
  contentType="text/html;charset=GBK"
  isErrorPage="true"
  session="true"
%>
<html>
<head>
    <title>错误</title>
</head>
<body style="color:red;"><br><br><br><br>
    <center>
        <%// 收取错误信息并输出到网页给用户看
            String sError=request.getParameter("msg");
            out.println("<br>");
            out.println(sError);
        %>
        <hr></hr>
    </ccnter>
</body>
</html>
```

Servlet 会将具体的错误信息作为参数传给错误页面,错误页负责将收到的错误信息进行显示。

相册首页使用的 frameset 对整个窗口进行分帧处理。albumindex.jsp 的代码如下:

```jsp
<%@ page
 contentType="text/html;charset=GBK"
%>
<html>
    <head>
        <title>我的相册</title>
        <meta http-equiv="Content-Type"
                content="text/html; charset=gb2312">
    </head>
    <frameset rows="15%,*">
    <frame name="topFrame" scrolling="NO" noresize
        src="" frameborder="0">
    <frameset cols="15%,*">
    <frame name="leftFrame" scrolling="NO" noresize
        src="accountMenu.jsp" frameborder="0">
    <frame name="mainFrame" src="myalbum.jsp" frameborder="0">
    </frameset>
    </frameset>
</noframes><body>
<p>Sorry!This page uses frames,but your browser doesn't support
```

```
them.</p>
   </body></noframes>
   </html>
```

这里将右边的窗格默认设置为显示 myalbum.jsp（我的相册）页面。myalbum.jsp 的代码如下：

```
<%@ page
 contentType="text/html;charset=GBK"
 import="java.util.*,com.javabean.AlbumInfo"
 session="true"
%>
<html>
<head>
   <title></title>
   <style type="text/css">
   .newjoyo_vcd1 {
     border-bottom:1px ridge #C49238;border-left:1px ridge #C49238;
     border-right:1px ridge #C49238;border-top:1px ridge #C49238;
     line-height:18px;
   }
   </style>
</head>
<body>
<center>
   <form action="AlbumServlet" method="post">
   <table border="0" width="90%" height="90%"
          style="border-color:#C49238"
         class="newjoyo_vcd1" cellpadding="0" cellspacing="0"
         bgcolor="azure">
   <tr><td height="10"></td></tr>
   <tr align="center">
   <%
   Vector vec=(Vector)session.getAttribute("userAlbum");
   for(Enumeration enu=vec.elements();enu.hasMoreElements();) {
          AlbumInfo ai=(AlbumInfo)enu.nextElement();
          if(vec.size()%5==0) {
                out.println("<br>");
          }
   %>
   <td align="center">
          <table border="0" width="15%" height="20%"
                style="border-color:#000000"
                class="newjoyo_vcd1" cellpadding="0"
```

```
cellspacing="0">
            <tr height="100" align="center"><td><img src="image/
                <%= ai.getAlbumTypeId() %>.jpg" width="110"
height="100">
            </td></tr>
            <tr height="20" align="center"><td>
                <a href="AlbumServlet?action=photo&albumId=<%=
ai.getAlbumId() %>"
                    target="_blank"><%= ai.getAlbumName() %></a></td></
tr>
            </table>
        </td>
        <%
        }
        %>
        </tr></table></form></center>
    </body>
    </html>
```

左边窗格列出了用户可以进行的操作。accountMenu.jsp 的代码如下：

```
<%@ page
 contentType="text/html;charset=GBK"
%>
<LINK href="css/default.css" type=text/css rel=stylesheet>
<LINK href="css/06default.css" type=text/css rel=stylesheet>
<html>
<head><title></title></head>
<body>
  <center><br><br>
  <table cellSpacing=8 cellPadding=0 border=0 bgcolor="azure">
  <tbody>
   <tr><td>
      <a href="AlbumServlet?action=myAlbum" target="mainFrame">我的相册
</a>
      </td></tr><tr>
      <td>
        <a href="AlbumServlet?action=newAlbum" target="mainFrame">新建相
册 </a>
      </td></tr><tr>
      <td>
        <a href="AlbumServlet?action=uploadPhoto" target="mainFrame">上
传照片 </a>
      </td></tr><tr>
```

```
        <td>
            <a href="AlbumServlet?action=deleteAlbum" target="mainFrame">删除相册</a>
        </td></tr><tr>
        <td>
            <a href="AlbumServlet?action=LogOut" target="_parent">退出登录</a>
        </td></tr></tbody></table></center></body>
</html>
```

用户可以在此页面选择要进行的操作，然后会将 action 分数提交到服务器由 Servlet 进行处理。比如，用户可以在相册首页选择已有的相册并查看相册中的照片。

2. 照片显示模块

在本应用中，当用户单击我的相册或相册首页中的相册名称链接时，都会涉及图片的显示问题。用户在创建新的相册时，系统会根据用户所选择的相册类型来决定相应相册类型的默认外观图片（如图 12-3 所示）。用户可以单击选择某个相册，请求被提交到服务器后由 Servlet 处理，照片显示模块的 Servlet 代码：

```
if(action.equals("photo")) {
            // 收取用户提交的参数
            String albumId=request.getParameter("albumId");
            // 调用 JavaBean 从数据库提取信息
            Vector vec=AlbumDAOBean.getAlbumPhotoInfo(albumId);
            // 将结果存入 request 对象中
            request.setAttribute("photoVec",vec);
            this.forward(request,response,"/photo.jsp");
```

然后会将此相册中的照片显示出来给用户看。photo.jsp 的代码如下：

```
<%@ page
 contentType="text/html;charset=GBK"
 import="java.util.*,com.javabean.Photo"
 session="true"
%>
<html>
<head>
    <title>相册明细</title>
    <style type="text/css">
    .newjoyo_vcd1 {
        border-bottom:1px ridge #C49238;border-left:1px ridge #C49238;
        border-right:1px ridge #C49238;border-top:1px ridge #C49238;
        line-height:18px;
    }
```

```
        </style>
    </head>
    <body>
    <center>
        <form action="AlbumServlet" method="post">
        <table border="0" width="90%" height="90%"
                style="border-color:#C49238"
            class="newjoyo_vcd1" cellpadding="0" cellspacing="0"
            bgcolor="azure">
        <tr><td height="10"></td></tr>
        <tr align="center">
        <%// 从 request 获取参数
            Vector vec=(Vector)request.getAttribute("photoVec");
    // 将内容输出到网页
            for(Enumeration enu=vec.elements();enu.hasMoreElements();) {
                Photo photo=(Photo)enu.nextElement();
                if(vec.size()%5==0) {
                    out.println("<br>");
                }
        %>
            <td align="center">
                <table border="0" width="15%" height="20%"
                        style="border-color:#000000" class="newjoyo_vcd1"
                        cellpadding="0" cellspacing="0">
                <tr height="80" align="center"><td>
        <%
         String s=photo.getPhotoId();
         if (true)
         {
              out.println("<img height=122 width=130 src=\""
              +request.getContextPath()+"/PhotoServlet?photoId="
+s+"\">");
         }
        %></td></tr>
                <tr height="20" align="center"><td>
                <a href="PhotoServlet?photoId=
                    <%= photo.getPhotoId() %>" title="单击查看大图 ">
                    <%= photo.getPhotoName() %></a>
                <a href="AlbumServlet?action=delPhoto&photoId=
                    <%= photo.getPhotoId() %>&albumId=<%= photo.getAlbumId() %>"
                    title=" 单击删除此图片 "> 删除 </a> 
                </td></tr></table></td>
```

```
        <%
            }
        %>
        </tr></table></form></center>
    </body>
</html>
```

从上面的 JSP 代码可以看出,在显示照片时用到了另一个 Servlet(PhotoServlet),这个 Servlet 通过照片编号从数据库中提取出照片并将图片的二进制数组表示写入输出流来显示图片。PhotoServlet 的代码如下:

```
import javax.servlet.*;
import javax.servlet.http.*;
import java.io.*;
import java.util.*;
import com.album.AlbumDAOBean;
import com.javabean.AlbumPhoto;
public class PhotoServlet extends HttpServlet {
    // 处理 POST 请求
    public void doPost(HttpServletRequest request,HttpServletResponse response)
                                                          throws ServletException,IOException {
        response.setContentType("image/jpeg");
        String photoId=request.getParameter("photoId");
        byte[] bi=null;
        OutputStream os=null;
        // 获取图片
        Vector vec=AlbumDAOBean.getPhoto(photoId);

        for(Enumeration enu=vec.elements();enu.hasMoreElements();) {

            AlbumPhoto ap=(AlbumPhoto)enu.nextElement();
            bi=ap.getPhotoByte();
        }
        try {// 输出图片
            os=response.getOutputStream();
            os.write(bi);
            os.flush();
        } catch (IOException ie) {
            ie.printStackTrace();
        } finally {
            os.close();
        }
```

```
        }
    // 处理 GET 请求
    public void doGet(HttpServletRequest request,HttpServletResponse response)
                                                                    throws ServletException,IOException {
        this.doPost(request,response);
    }
}
```

请注意这里的 response.setContentType("image/jpeg");语句，当输出的内容为图片时，必须将响应的 ContentType 设置为 image/jpeg，然后通过 HttpServletResponse 对象上的 getOutputStream() 方法获取 OutputStream（输出流）对象，并调用 OutputStream 对象上的 write() 方法将图片的字节数组写入输出流，最后一定要使用 OutputStream 对象上的 flush() 方法，否则图片显示可能会不正常。运行后得到如图 12-14 所示的页面。

图 12-14 相册照片明细

在此页面中，用户可以单击图片的名称查看原始图片，也可以单击"删除"链接删除图片。删除操作的 Servlet 代码如下：

```
if(action.equals("delPhoto")) {
    // 如果用户执行的是删除照片操作，就执行此段代码
            String photoId=request.getParameter("photoId");
            // 调用 JavaBean 从数据库中删除照片
            boolean b=AlbumDAOBean.deletePhoto(photoId);
            if(b) {// 如果删除成功，就将现有数据从数据库中提取出来返回给用户
                String albumId=request.getParameter("albumId");
                Vector vec=AlbumDAOBean.getAlbumPhotoInfo(albumId);
                request.setAttribute("photoVec",vec);
                this.forward(request,response,"/photo.jsp");
            } else {
                this.forward(request,response,"/error.jsp?msg=
```

删除失败！");
                }
            }

当用户提交的 action 值为 delPhoto 时执行以上操作，然后返回 photo.jsp 显示用户删除后的其他照片。

3. 用户上传照片模块

用户上传照片模块主要用于实现用户从本地将照片上传到服务器并保存到数据库的相应相册中。下面是用户上传照片模块的 JSP（uploadPhoto.jsp）代码如下：

```jsp
<%@ page
 contentType="text/html;charset=GBK"
 import="java.util.*,com.javabean.AlbumInfo"
 errorPage="error.jsp"
 session="true"
%>
<html>
    <head><title></title></head>
    <body><br><br><br><br><br>
    <form action="AlbumServlet" method="post" name="updateImageForm2">
    <center><table><tbody>
    <tr>
    <td><p>选择相册：
        <select name="albumId">
    <%
        Vector vec=(Vector)session.getAttribute("userAlbum");
        for(Enumeration enu=vec.elements();enu.hasMoreElements();) {
            AlbumInfo ai=(AlbumInfo)enu.nextElement();
            if(vec.size()%5==0) {
                out.println("<br>");
            }
    %>
    <option value=<%= ai.getAlbumId() %>><%= ai.getAlbumName() %></option>
    <%
        }
    %></select><br></p></td></tr>
    <tr><td>
         <p>
      照片名称：
      <input type="test" name="photoName" size="30" maxlength="80"><br>
    </p></td></tr>
    <tr><td>
```

```html
     <p>
         路      径：
       <input type="file" name="path" size="30" maxlength="80"><br>
     </p></td></tr>
     <tr><td height="10"></td></tr>
     <tr align="center"><td>
         <p>
          <input type="hidden" name="action" value="addImageSubmit">
          <input type="submit" value=" 上传照片 "></p>
     </td></tr>
     </tbody></table></center></form>
    </body>
</html>
```

运行得到如图 12-15 所示的页面。

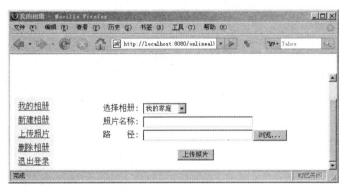

图 12-15 上传照片页面

在图 12-15 所示的页面中，用户可以选择将上传的照片保存到已有的相册中。如果选择"我的家庭"，那么上传的图片就会被保存到用户创建的名称为"我的家庭"的相册中，然后填写照片的名称并选择本地照片的路径，单击"上传照片"按钮，请求将被发送到服务器端由 Servlet 进行处理。Servlet 中此模块的代码如下：

```java
if (action.equals("uploadPhoto")) {
                this.forward(request,response,"/uploadPhoto.jsp");
} else if (action.equals("addImageSubmit")) {
                // 如果用户执行的是上传照片的操作，就执行此段代码
                // 收取用户提交的参数
                String albumId=request.getParameter("albumId");
                String fileName=request.getParameter("path").trim();
                String photoName=request.getParameter("photoName");
                String stemp=Long.valueOf(System.currentTimeMillis()).
toString();
```

```
                    String photoId=stemp.substring(6,stemp.length());
                    // 调用JavaBean将图片存入数据库中
                    boolean b=AlbumDAOBean.insertImage(albumId,photoId,pho
toName,fileName);
                    if(b) {
                            this.forward(request,response,"/myalbum.jsp");
                    } else {
                            this.forward(request,response,"/error.jsp?msg=
上传失败，请重试！");
                    }
                }
```

如果上传成功，就会跳转到 myalbum.jsp（我的相册）页面显示，否则就跳转到 error.jsp（错误页）显示错误信息。

4. 新建相册模块

用户初次注册后，系统并没有提供默认的相册。这时用户就需要建立自己的相册，然后才能上传照片到服务器。下面将介绍如何实现创建相册模块。用户登录成功后，可以在相册首页单击"新建相册"链接到新建相册页面，newAlbum.jsp 的代码如下：

```
<%@ page
 contentType="text/html;charset=GBK"
 import="java.util.*"
 errorPage="error.jsp"
 session="true"
%>
<html>
  <head><title></title></head>
  <body><br><br><br><br>
  <%
        HashMap hm=(HashMap)request.getAttribute("albumTyteInfo");
  %>
  <form action="AlbumServlet" method="post" name="updateImageForm2">
  <center><table><tbody>
  <tr><td><p>
        相册类型：
      <select name="typeId">
  <%
    Set ks=hm.keySet();
    for(Iterator ii=ks.iterator();ii.hasNext();) {
            String id=(String)ii.next();
            String name=(String)hm.get(id);
  %>
      <option value=<%= id %>><%= name %></option>
```

```
                <%
            }
%> </select><br></p></td></tr><tr><td height="10"></td></tr>
<tr><td>
        <p>
    相册名称：
    <input type="test" name="albumName" size="30" maxlength="80"><br>
    </p></td></tr><tr><td height="10">
</td></tr>
<tr align="center"><td>
        <p>
        <input type="hidden" name="action" value="newAlbumSubmit">
        <input type="submit" value=" 创建相册 "></p>
</td></tr>
</tbody></table></center></form>
</body>
</html>
```

运行结果如图 12-16 所示。

图 12-16 新建相册页面

在如图 12-16 所示的页面中，用户可以选择新建相册的类型，如"旅游见闻"。这样系统就会自动根据用户所选的相册类型给该相册一张默认的外观图片，然后输入相册名称，单击"创建相册"按钮，请求将被提交到服务器端由 Servlet 进行处理。Servlet 中此模块的代码如下：

```
if(action.equals("newAlbum")) {
            HashMap hm=AlbumDAOBean.getAlbumTyteInfo();
            request.setAttribute("albumTyteInfo",hm);
            this.forward(request,response,"/newAlbum.jsp");
    } else if(action.equals("newAlbumSubmit")) {// 新建相册
            String albumName=request.getParameter("albumName");
            if(albumName==null||albumName.equals("")) {
                this.forward(request,response,"/error.jsp?msg=
```

```
相册名称不能为空！");
                        return;
                    }
                    String uaId=(String)session.getAttribute("uaId");
                    String stemp=Long.valueOf(System.currentTimeMillis()).
toString();
                    String albumId=stemp.substring(5,stemp.length());
                    String userName=(String)session.
getAttribute("userName");
                    String typeId=request.getParameter("typeId");
                    // 将新建相册信息插入数据库
                    boolean b=AlbumDAOBean.newAlbum(uaId,albumId,typeId,al
bumName);
                    if(b) {
                    // 如果新建成功，就将现有相册信息从数据库中提取出来给用户显示
                        Vector vec=AlbumDAOBean.getAlbumInfo(userName);
                        session.setAttribute("userAlbum",vec);
                        this.forward(request,response,"/myalbum.jsp");
                    } else {
                        this.forward(request,response,"/error.jsp?msg=
新建相册错误！");
                    }
                }
```

如果相册创建成功，就会跳转到 myalbum.jsp（我的相册）页面显示，否则就跳转到 error.jsp（错误页）显示错误信息。

```
String stemp=Long.valueOf(System.currentTimeMillis()).toString();
String albumId=stemp.substring(5,stemp.length());
```

上面一段代码的功能是为所建相册随机分配一个相册编号。

5. 删除相册模块

用户也可以将所创建的相册删除，只要单击"删除相册"链接，请求就会被送到服务器由 Servlet 进行处理。Servlet 中此模块的代码如下：

```
if(action.equals("deleteAlbum")) {
                    // 如果用户执行的是删除相册的操作，就执行此段代码
                    String userName=(String)session.
getAttribute("userName");
                    Vector vec=AlbumDAOBean.getAlbumInfo(userName);
                    request.setAttribute("albumInfo",vec);
                    this.forward(request,response,"/deleteAlbum.jsp");
```

第 12 章　Java 项目开发

```
        }
```

完成上面的操作后将跳转到 deleteAlbum.jsp 页面，此页面列出了用户当前所有相册以及各个相册的基本信息（相册名称、相册编号、创建相册的时间、相册类型和当前相册中的照片数量）。deleteAlbum.jsp 的代码如下：

```jsp
<%@ page
 contentType="text/html;charset=GBK"
 import="java.util.*,com.javabean.AlbumInfo"
 errorPage="error.jsp"
 session="true"
%>
<html>
<head><title></title>
    <style type="text/css">
        .newjoyo_vcd1 {
            border-bottom:1px ridge #C49238;border-left: 1px ridge #C49238;
            border-right:1px ridge #C49238;border-top: 1px ridge #C49238;
            line-height:18px;
        }
    </style>
</head>
<body><center>
    <form action="AlbumServlet" method="post">
    <table border="0" width="90%" height="20%" style="border-color:#C49238"
            class="newjoyo_vcd1" cellpadding="0" cellspacing="0"
            bgcolor="azure">
    <tr><td height="10"></td></tr>
    <tr align="center" bgcolor="#C0C0C0">
            <th> 编号 </th>
            <th> 相册名称 </th>
            <th> 创建时间 </th>
            <th> 相册类型 </th>
            <th> 照片数量 </th>
            <th> 删除 </th></tr>
    <%
            Vector vec=(Vector)request.getAttribute("albumInfo");
            for(Enumeration enu=vec.elements();enu.hasMoreElements();) {
                AlbumInfo ai=(AlbumInfo)enu.nextElement();
    %>
            <tr align="center">
            <td align="center"><%= ai.getAlbumId() %></td>
            <td align="center"><%= ai.getAlbumName() %></td>
            <td align="center"><%= ai.getNewTime() %></td>
```

```
                <td align="center"><%= ai.getAlbumType() %></td>
                <td align="center"><%= ai.getPhotoCount() %></td>
                <td align="center">
                <a href="AlbumServlet?action=del&albumId=<%= ai.getAlbumId() %>">
                                   删除</a></td>
            </tr>
            <%}
        %>
        </table></form></center>
</body>
</html>
```

运行得到如图 12-17 所示的页面。

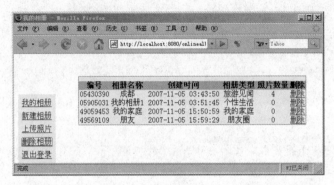

图 12-17 删除相册页面

12.5 要点总结

本章通过三个综合实例着重介绍了程序设计开发的完整过程。一个从需求、分析设计、编码实现到测试的全过程，其中重点突出了设计实现时的思维过程。

12.6 编程练习

根据 Java 开发流程开发一个学生信息管理系统，满足学生基本信息维护、学生成绩维护等功能。